ARTIFICIAL INTELLIGENCE

Ms. B. Bhuvaneswari
Ms. E. M. Roopadevi
Dr. S. Anandamurugan

2021
Associated Publishing Company®
A Division of
Astral International Pvt. Ltd.
New Delhi – 110 002

ISBN: 9789390371792 (Int. Edition)

Published by : **Associated Publishing Company®**
A Division of
Astral International Pvt. Ltd.
– ISO 9001:2015 Certified Company –
4736/23, Ansari Road, Darya Ganj
New Delhi-110 002
Ph. 011-43549197, 23278134
E-mail: info@astralint.com
Website: www.astralint.com

Acknowledgement

In the first place, we would like to express our sincere gratitude to Thiru. P. Sachithanandan, Correspondent, Kongu Engineering College, for having provided us with all the necessary facilities to undertake this work.

We feel extremely grateful for the strong support received from ***Dr. V. Balusamy, Principal*** and ***Dr. S. Varadhaganapathy***, Head, Department, Information Technology, Kongu Engineering College for their commendable help and support during this work.

We extend our sincere thanks to all our colleagues for their comments and suggestions. Finally, we would like to thank everyone who encouraged us in every step of the way in the successful completion of the book.

Ms. B. Bhuvaneswari
Ms. E. M. Roopadevi
Dr. S. Anandamurugan

Preface

This book is intended for use by new students taking their course in Artificial Intelligence. AI is the study of the design of intelligent computational agents. The book is structured as a textbook but it is designed to be accessible to a wide audience. This book uses simple approach for better understanding with the help of diagrams. This book can be used as an introductory text on artificial intelligence for advanced undergraduate or graduate students in computer science or related disciplines such as computer engineering, computer technology and information technology.

Ms. B. Bhuvaneswari
Ms. E. M. Roopadevi
Dr. S. Anandamurugan

CONTENTS

1

Introduction to AI and Production Systems

1.1. Introduction to AI

Artificial Intelligence (AI) is a branch of Science which deals with helping machines finds solutions to complex problems in a more human-like fashion. This generally involves borrowing characteristics from human intelligence, and applying them as algorithms in a computer friendly way.

A more or less flexible or efficient approach can be taken depending on the requirements established, which influences how artificial the intelligent behaviour appears. From a business perspective AI is a set of very powerful tools, and methodologies for using those tools to solve business problems.

- ✯ From a programming perspective, AI includes the study of symbolic programming, problem solving, and search.
  - Typically AI programs focus on symbols rather than numeric processing.
  - Problem solving - achieve goals.
  - Search - seldom access a solution directly. Search may include a variety of techniques.

1.1.1. Definition

How to make computers do things at which, at the moment, people are better.

What is Artificial Intelligence?

- ✯ Depends on your perspective:
  - Philosophical: a method for modeling intelligence.
  - Psychological: a method for studying human cognition.
  - Engineering: a set of methodologies for building complex programs.

1.1.2. Categories

- ✯ System that think like humans
- ✯ System that act like humans
- ✯ System that think rationally
- ✯ System that act rationally

Relationships of AI to other Fields

- ✯ Economics – rational decision-making.
- ✯ Mathematics – logic and number theory.
- ✯ Neuroscience – neural networks.

1.1.3. The Future of AI

- ✯ It is very unlikely that the "traditional" approach to strong AI will achieve its long-term goal.
- ✯ It is unlikely that the "traditional" approach will produce large-scale (as opposed to niche) breakthroughs in its way to (not) achieving its long-term goal.
- ✯ It is very likely that weak AI will achieve its short-term goals (more specifically, building intelligent environments), also via a strong synergy with computer science.

1.2. Problem Formulation & Problem Definition

Problem formulation means choosing a relevant set of states to consider, and a feasible set of operators for moving from one state to another.

Search is the process of considering various possible sequences of operators applied to the initial state, and finding out a sequence which culminates in a goal state.

A search problem consists of the following:

- ✯ S: the full set of states
- ✯ s0 : the initial state
- ✯ A : S→S is a set of operators
- ✯ G is the set of final states. Note that GÍS

The search problem is to find a sequence of actions which transforms the agent from the initial state to a goal state gÎG.

A simple search procedure is:

Do until a solution is found or the state space is exhausted.

1. Check the current state.
2. Execute allowable actions to find the successor states.
3. Pick one of the new states.
4. Check if the new state is a solution state.

If it is not, the new state becomes the current state and the process is repeated

Basic Search Algorithm is:

Let L be a list containing the initial state (L= the fringe).

Loop if L is empty return failure.

Node <- select (L).

If Node is a goal then return Node (the path from initial state to Node) else generate all successors of Node, and merge the newly generated states into L End Loop

1.3. Production Systems

Production system is a mechanism that describes and performs the search process.

A production system consists of four basic components:

1. A set of rules of the form Ci → Ai where Ci is the condition part and Ai is the action part. The condition determines when a given rule is applied, and the action determines what happens when it is applied (*i.e.*, A set of rules, each consist of left side (a pattern) that determines the applicability of the rule and a right side that describes the operations to be performed if the rule is applied).

2. One or more knowledge databases that contain whatever information is relevant for the given problem. Some parts of the database may be permanent, while others may temporary and only exist during the solution of the current problem. The information in the databases may be structured in any appropriate manner.

3. A control strategy that determines the order in which the rules are applied to the database, and provides a way of resolving any conflicts that can arise when several rules match at once.

A production system classifies the common characteristics shared by a class of reasoning problems, and has the following components:

1. Collection of states
   - ★ Start or initial state.
   - ★ Goal state.
2. Collection of productions: rules or moves
   - ★ Each production may have preconditions.
3. Control system: decides which production to apply next.

1.3.1. Data Processing for Production Systems

- ★ State graph = states, productions, and preconditions.
- ★ A graph consists of nodes and arcs(arrows) connecting nodes.
- ★ A state graph has nodes representing states and arrows representing rules.
- ★ The arc linking two nodes signifies two states can be shift to each other using the rule; the absence of arcs implicitly indicates the preconditions are not met.
- ★ The problem magnitude may be too large for explicitly showing the entire state graph.
- ★ Partial representation of the state graph will help understand the problem.

1.4. Control Strategies

By considering control strategies we can decide which rule to apply next during the process of searching for a solution to problem.

The two requirements of good control strategy are that

- ★ **It should cause motion**: consider water jug problem, if we implement control strategy of starting each time at the top of the list of rules, it will never leads to solution. So we need to consider control strategy that leads to solution.
- ★ **It should be systematic:** choose at random from among the applicable rules. This strategy is better than the first. It causes the motion. It will lead to the solution eventually. Doing like this is not a systematic approach and it leads to useless sequence of operators several times before finding final solution.

1.4.1. Systematic Control Strategy for the Water Jug Problem

1.4.1.1. Breadth First Search (Blind Search)

Let us discuss these strategies using water jug problem. These may be applied to any search problem.

- ★ Construct a tree with the initial state as its root.
- ★ Generate all the offspring of the root by applying each of the applicable rules to the initial state.

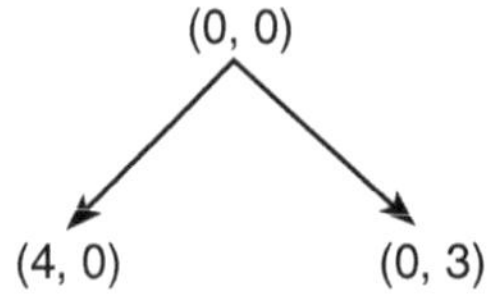

- ★ Now for each leaf node, generate all its successors by applying all the rules that are appropriate.

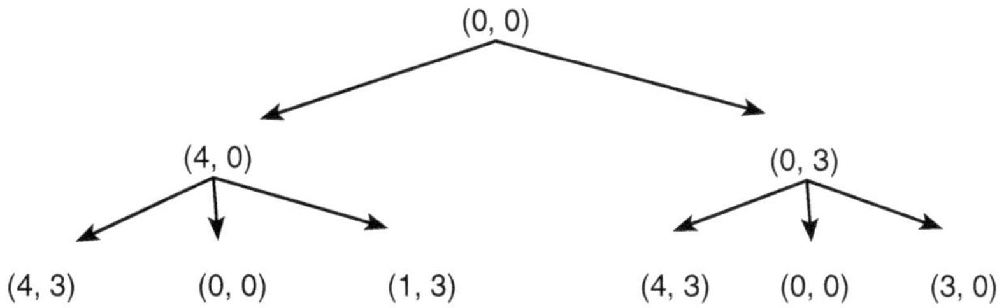

- ✯ Continue this process until some rule produces a goal state.

Algorithm

1. Create a variable called NODE-LIST and set it to initial state.
2. CHAPTER a goal state is found or NODE-LIST is empty do
   a. Remove the first element from NODE-LIST and call it E. if NODE-LIST is empty, quit.
   b. For each way that each rule can match the state described in E do:
      i. Apply the rule to generate a new state.
      ii. If the new state is a goal state, quit and return this state.
      iii. Otherwise, add the new state to the end of NODE-LIST.

1.4.1.2. Depth First Search

Algorithm

1. If the initial state is the goal state, quit and return success.
2. Otherwise do the following until success or failure is signaled:
   a. Generate a successor, E, of initial state. If there are no more successor, signal failure.
   b. Call depth first search, with E as the initial state.
   c. If the success is returned, signal success. Otherwise continue in this loop.

Backtracking

- ✯ In this search, we pursue a single branch of the tree until it yields a solution or until a decision to terminate the path is made.
- ✯ It makes sense to terminate a path if it reaches dead-end, produces a previous state. In such a state backtracking occurs.

Chronological Backtracking: order in which steps are undone depends only on the temporal sequence in which steps were initially made.

- ✯ Specifically most recent step is always the first to be undone. This is also simple backtracking.

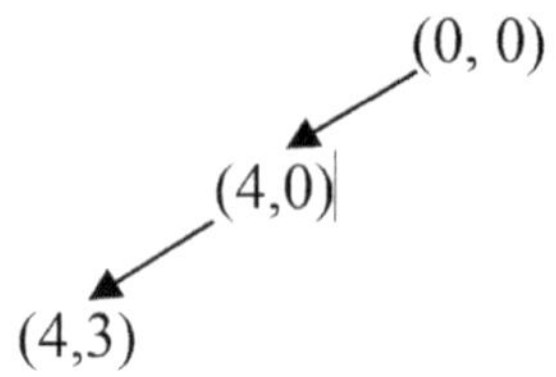

Advantages of Depth First Search

* DFS requires less memory since only the nodes on the current path are stored.
* By chance DFS may find a solution without examining much of the search space at all.

Advantages of Breath First Search

* BFS cannot be trapped exploring a blind alley.
* If there is a solution, BFS is guaranteed to find it.
* If there are multiple solutions, then a minimal solution will be found.

1.5. Search Strategies

Formulate a problem as a state space search by showing the legal problem states, the legal operators, and the initial and goal states.

* A state is defined by the specification of the values of all attributes of interest in the world.
* An operator changes one state into the other; it has a precondition which is the value of certain attributes prior to the application of the operator, and a set of effects, which are the attributes altered by the operator.
* The initial state is where you start.
* The goal state is the partial description of the solution.

State Space Search Notations

The set of notations involved in the state space search is :

1. An initial state is the description of the starting configuration of the agent
2. An action or an operator takes the agent from one state to another state which is called a successor state. A state can have a number of successor states.
3. A plan is a sequence of actions. The cost of a plan is referred to as the path cost.

1.5.1. Search Strategies

* Completeness
* Optimality
* Complexity

- ☆ Possibility to backtrack
- ☆ Informedness
  - A strategy is defined by picking the order of node expansion.
  - Time and space complexity are measured in terms of,

b max. branching factor of the search tree.

→

d depth of the least_cost solution.

→

m max. depth of the state space.

→

1.5.2. Informed Search Strategies

- ☆ Relies on additional knowledge about the problem or domain.
- ☆ Used to distinguish more promising path to words a goal.
- ☆ Performs much better than uninformed search.

Best First Search

- ☆ Relies on an evaluation function that gives an indication of how useful it would be to expand node.
- ☆ The node with the lowest value is expanded first.

Greedy Best First Search

- ☆ Minimizes the estimated cost to a goal.
- ☆ Expand the node that seems to be closest to a goal.
- ☆ Utilizes a heuristic function as evaluation function.
- ☆ Often better than depath first, at though worst time complexities are equal or words.

A* Search

- ☆ Combines greedy and uniform cost search to find the cheapest path through the current node.
- ☆ $f(n)=g(n)+h(n)$.
- ☆ Heuristics must be admissible.
- ☆ Very good search method, but with complexity.

Memory – Bounded Search

- ☆ Search algorithms that try to conserve memory.
- ☆ Iterative depending A*.
- ☆ Simplified memory – bounded A*.

Types:

1. Uninformed search strategies.
2. Informed search strategies.

1.5.3. Uninformed Search Strategies

- ★ Search has no additional information about the states beyond that provided in the problem definition.
- ★ Less effective in search method.

Strategies

- ★ Breadth first search
- ★ Uniform cost search
- ★ Depth first search
- ★ Depth limited search
- ★ Iterative deepening DFS
- ★ Bidirectional search

1.BFS

- ★ Root node is expanded first,then all the successors of the root node are expanded next.
- ★ All nodes at a given depth are expanded in the search tree.

Implementation: (FIFO) Queue

- ★ Nodes visited first will be expanded first.

DFS

- ★ Expand deepest unexpanded node.
- ★ LIFO queue.
- ★ Continuous exploring newly generated nodes.
- ★ Tree-search method.

Uniform Cost Search

- ★ Nodes with the lowest cost are explored first.
- ★ Similarly to BFS.
- ★ G(n) = path cost(n) = sum of individual edge costs to reach the current node.

Depth Limited Search

- ★ Similarly to DFS (but with a limit).
- ★ Overcomes problems with infinite paths.
- ★ Sometimes a depth limit can be inferred or estimated from the problem description.
- ★ Must keep track of the depth.

Iterative Deepening

- Applies limited - depth with increasing depth limits.
- Combines advantages of BFS and DFS.
- Many states are expanded multiple times.
- One of the best uninformed search methods.

Bidirectional

- Search simultaneously from two directions.
  - → Forward from the initial and backward from the goal state.
- Lead to substantial savings if it is applicable.
- Severe limitations, predecessors must generated, which is not always possible.
- Search must be coordinated between the two searches.
- One search must keep all nodes in memory.

Heuristic Search Strategies

- To solve large problems with large number of possible states, problem-specific knowledge needs to be added to increase the efficiency of search algorithms.

Heuristic Evaluation Functions

- They calculate the cost of optimal path between two states. A heuristic function for sliding-tiles games is computed by counting number of moves that each tile makes from its goal state and adding these number of moves for all tiles.

A* SEARCH

It is best-known form of Best First search. It avoids expanding paths that are already expensive, but expands most promising paths first.

f(n) = g(n) + h(n), where

g(n) the cost (so far) to reach the node

h(n) estimated cost to get from the node to the goal

f(n) estimated total cost of path through n to goal. It is implemented using priority queue by increasing f(n).

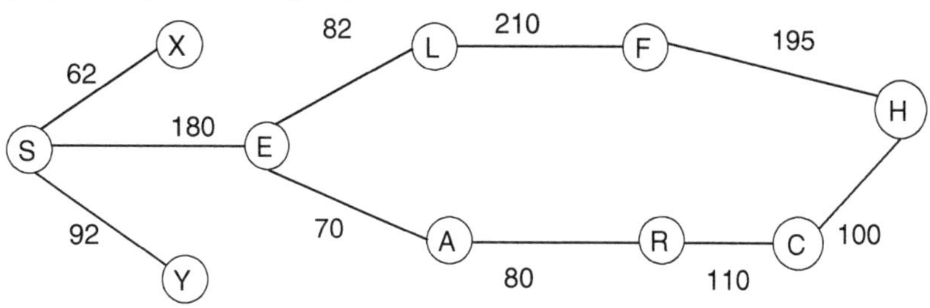

1.6. Problem Characteristics

Heuristic search is a very general method applicable to a large class of problem. In order to choose the most appropriate method (or combination of methods) for a particular problem it is necessary to analyze the problem along several key dimensions.

1.6.1. Is the problem decomposable into a set of independent smaller sub-problems?

Example: Suppose we want to solve the problem of computing the integral of the following expression $\int(x^2 + 3x + \sin^2 x * \cos^2 x)\,dx$

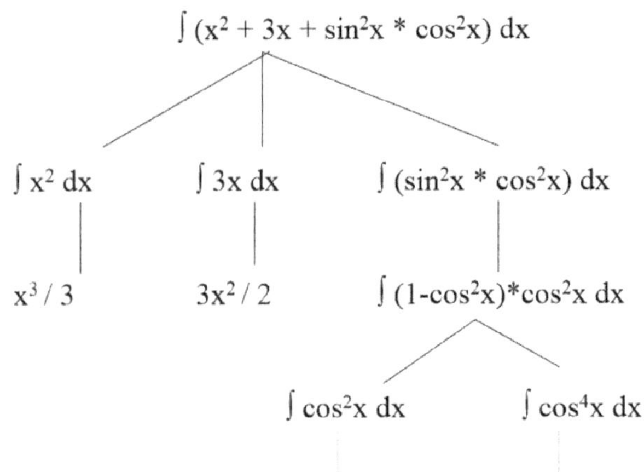

We can solve this problem by breaking it down into three smaller sub-problems, each of which we can then be solved using a small collection of specific rules.

- Decomposable problems can be solved by the divide and-conquer technique.
- Use of decomposing problems:
  - Each sub-problem is simpler to solve.
  - Each sub-problem can be handed over to a different processor. Thus can be solved in parallel processing environment.
- There are non-decomposable problems. For example, Block world problem is non-decomposable.

1.7. Production System Characteristics

Production systems are important in building intelligent matches which can provide us a good set of production rules, for solving problems.

There are four types of production system characteristics, namely:

1. Monotonic production system.
2. Non-monotonic production system.

3. Commutative law based production system, and lastly.
4. Partially commutative law based production system.

1. **Monotonic Production System (MPS):** The Monotonic production system (MPS) is a system in which the application of a rule never prevents later application of the another rule that could also have been applied at the time that the first rule was selected.
2. **Non-monotonic Production (NMPS):** The non-monotonic production system is a system in which the application of a rule prevents the later application of the another rule which may not have been applied at the time that the first rule was selected, *i.e.* it is a system in which the above rule is not true, *i.e.* the monotonic production system rule not true.
3. **Commutative Production System (CPS):** Commutative law based production systems is a system in which it satisfies both monotonic & partially commutative.
4. **Partially Commutative Production System (PCPS):** The partially commutative production system is a system with the property that if the application of those rules that is allowable & also transforms from state x to state _y'.

1.8. Rationality

A system is rational, if it does the "right thing".

- ✯ Acting humanly: Turing test approach

  e.g: Two persons and one machine.
- ✯ One person will be a questioner.
- ✯ Another person & machine will be a responder.
- ✯ Questioner aim is to determine which the person is & which the machine is.

Goal

Fool the questioner, believe that is the person. If a m/c succeeds at this, questioner will conclude that machine → acting humanly.

Capabilities

- ✯ Natural language processing
- ✯ Knowledge representations
- ✯ Automated reasoning
- ✯ Machine learning
- ✯ Total Turing test pass will need,

  Computer vision→to perceive objects

  Robotics→to manipulate and move them

Thinking Humanly : Cognitive modeling approach

- ✯ Machine program is constructed to think like a human, for that we need knowledge.
- ✯ Computer program→with input, output & time behaviour matches with a human behaviour.

  e.g: GPS

Think rationally: "Laws of the though" approach

Right thinking initiated the field called logic.

e.g: ("AI is a computer science subject ". All computer science subjects are easy, therefore easy).

Acting rationally: Rationally agent approach.

Agents operate under autonomous control and being capable of taking on another's goals.

Application of AI

- ✯ Gaming
- ✯ Journalism
- ✯ Media
- ✯ Finance
- ✯ Robotics
- ✯ Medical Diagnosis
- ✯ Quantum Science

1.9. Agents

- ✯ An agent is anything that can be viewed as perceiving its environment through sensors and acting upon that environment through actuators.
- ✯ Human agent: eyes, ears, and other organs for sensors; hands, legs, mouth, and other body parts for actuators.
- ✯ Robotic agent: cameras and infrared range finders for sensors; various motors for actuators.

Agents and Environments

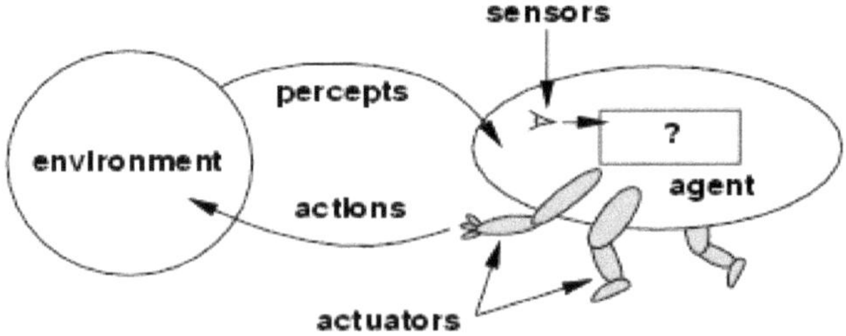

Agent Function

- ✯ The agent function maps from percept histories to actions: [f: P* → A].
- ✯ The agent program runs on the physical architecture to produce f agent = architecture + program.

1.9.1. Intelligent Agents

An agent that does the best job by understanding the environment is called intelligent agent. An agent always requires a certain amount of intelligence to perform its tasks. Consequently, one refers to intelligent agent.

1.9.1.1. Characteristics of Intelligent Agents

Internal Characteristics are Learning/ Reasoning

An agent has the ability to learn from previous experience and to successively adapt its own behavior to the environment.

Reactivity

An agent must be capable of reacting appropriately to influences or information from its environment.

Autonomy

An agent must have both control over its actions and internal states. The degree of the agent's autonomy can be specified. There may need intervention from the user only for important decisions.

Goal-oriented

An agent has well-defined goals and gradually influence its environment and so achieve its own goals.

External Characteristics are Communication

An agent often requires an interaction with its environment to fulfil its tasks, such as human, other agents, and arbitrary information sources.

Cooperation

Cooperation of several agents permits faster and better solutions for complex tasks that exceed the capabilities of a single agent.

Mobility

An agent may navigate within electronic communication networks.

Character

Like human, an agent may demonstrate an external behavior with many human characters as possible.

What are Software Agents Different from Traditional Software Programs?

Perception : Software agents are aware of environment changes.

Autonomy : Software agents perform tasks largely autonomously.

Learning : Software agents are accumulating knowledge (learning through experience)

Communication: Software agents communicate with users and/or other agents

1.10. Rational Agent

An agent should strive to "do the right thing", based on what it can perceive and the actions it can perform. The right action is the one that will cause the agent to be most successful

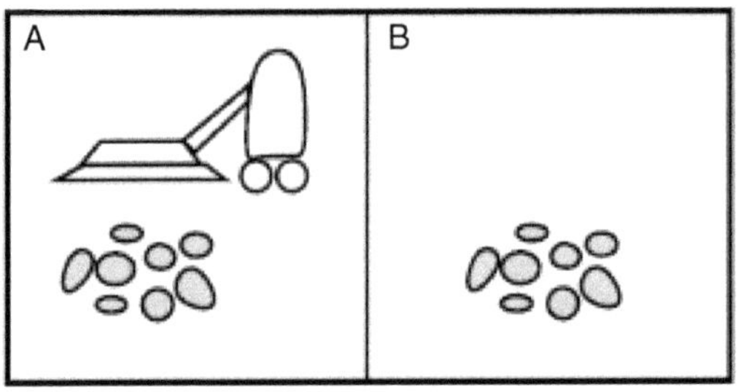

Vacuum Cleaner function

Performance measure: An objective criterion for success of an agent's behavior. For example: performance measure of a vacuum-cleaner agent could be amount of dirt cleaned up, amount of time taken, amount of electricity consumed, amount of noise generated, etc.

For each possible percept sequence, a rational agent should select an action that is expected to maximize its performance measure, given the evidence provided by the percept sequence and whatever built-in knowledge the agent has.

Precept

- Agents known inputs at any given instant are called percept.
- Percept sequence (Every thing agent perceived).
- Agent Function (Mathematical behaviour deals with agents behaviours maps to percept sequence into an action).
- Agent program (Agent function for an agent will be implement).

  e.g: Vacuum cleaner.

Rationality Depends on four Things

- Performance measure defines the criterion of success.
- Agent's prior knowledge of the environment.
- The actions that the agent can perform.
- The agent's percept sequence to date.

Good Behaviour

↓

Ration Agent →one does the right thing

↓

Success based on right action.

Measure the success (Performance measures).

→Agents generate sequence of action in an environment.

According to the percept receives.

e.g: Vacuum cleaner problem .

Clean up

Dump

Clean it up again.

IDEAL RATIONAL AGENT	IRRATIONAL AGENT
Select an action expected to maximum its performance measure.	Unnecessary action taken.

1.10.1. Omniscience

Known the actual outcome of its actions and can act accordingly.

e.g: A person walked on the street & met his old friend.

→ Rationality is not the same as perfection.

↓ ↓

Maximize expected performance   maximum actual performance

1.10.2. Learning

What agent perceives?

Agents initial configuration reflect some prior knowledge of the environment but the agent gains experience may be modified.

1.10.3. Autonomy

If the agents actions are based completely on built in knowledge rather than its own percepts, then the agents lacks autonomy.

A system is autonomous to the extent that its behaviour is determined by its own experience.

1.10.4. Nature of Environments

- ★ "Task environments" are the problems.
- ★ "Rational agents" are the solutions.

1.10.5. PEAS

- ★ PEAS: Performance measure, Environment, Actuators, Sensors.
- ★ Must first specify the setting for intelligent agent design.

Consider, *e.g.*, the task of designing an automated taxi driver:

Performance measure	:	Safe, fast, legal, comfortable trip, maximize profits.
Environment	:	Roads, other traffic, pedestrians, customers.
Actuators	:	Steering wheel, accelerator, brake, signal, horn.
Sensors	:	Cameras, sonar, speedometer, GPS, odometer, engine sensors, keyboard.

Example:

Agent	:	Medical diagnosis system.
Performance measure	:	Healthy patient, minimize costs, lawsuits.
Environment	:	Patient, hospital, staff.
Actuators	:	Screen display (questions, tests, diagnoses, treatments, referrals).
Sensors	:	Keyboard (entry of symptoms, findings, patient's answers).

1.10.6. Types of Environment

Fully Observable (vs. partially observable)

An agent's sensors give it access to the complete state of the environment at each point in time.

Deterministic (vs. stochastic)

The next state of the environment is completely determined by the current state and the action executed by the agent. (If the environment is deterministic except for the actions of other agents, then the environment is strategic).

Episodic (vs. sequential)

The agent's experience is divided into atomic "episodes" (each episode consists of the agent perceiving and then performing a single action), and the choice of action in each episode depends only on the episode itself.

Static (vs. dynamic)

The environment is unchanged while an agent is deliberating. (The environment is semidynamic if the environment itself does not change with the passage of time but the agent's performance score does).

Discrete (vs. continuous)

A limited number of distinct, clearly defined percepts and actions.

Single agent (vs. multiagent)

An agent operating by itself in an environment.

1.10.7. Agent types

* Four basic types in order of increasing generality:
  1. Simple reflex agents
  2. Model-based reflex agents
  3. Goal-based agents
  4. Utility-based agents

1. Simple Reflex Agents

Select actions on the basis of the current precept, by ignoring the rest of the precept history.

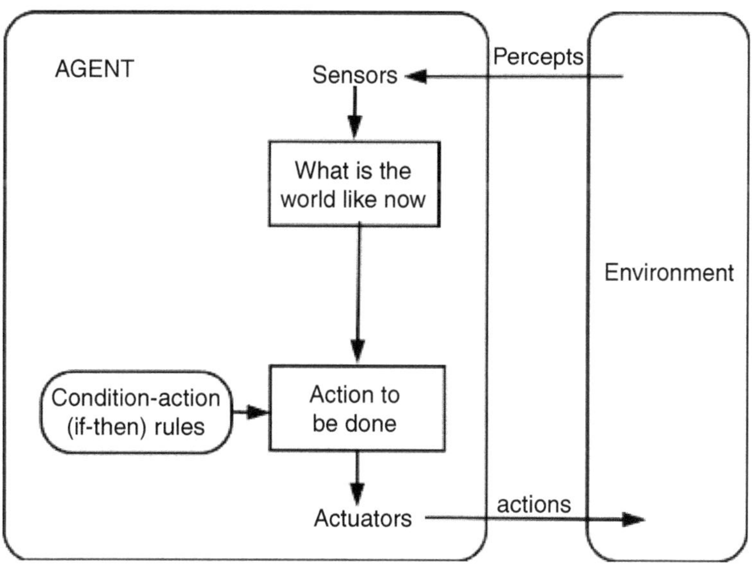

e.g: Vacuum Agent.

Decision is based only on current location.

Function Reflex Vacuum – Agent (location, status) returns an action.

if status = dirty, then return suck

else if location = A, then return right

else if location = B, then return left

e.g: Automated Taxi.

Car in front brakes and its brake lights come on initiate braking.

if car- in front- is –braking(condition)

then initiate – braking (action)

Function

SIMPLE – REFLEX – AGENT (Precept) returns on action.

Static : rules, a set of condition – action rules.

State INTERPRET – INPUT (Precept)

Rule RULE – MATCH (State, rules)

Action RULE – ACTION [rule]

return action.

2. Model Reflex Agent

- ✯ Agent keeps track of the part of the world it can't see now.
- ✯ Maintain some internal state depends on the precept history.
- ✯ Partially observable state:
  - how the world evolves
  - what my actions do

3. Goal Based Agents

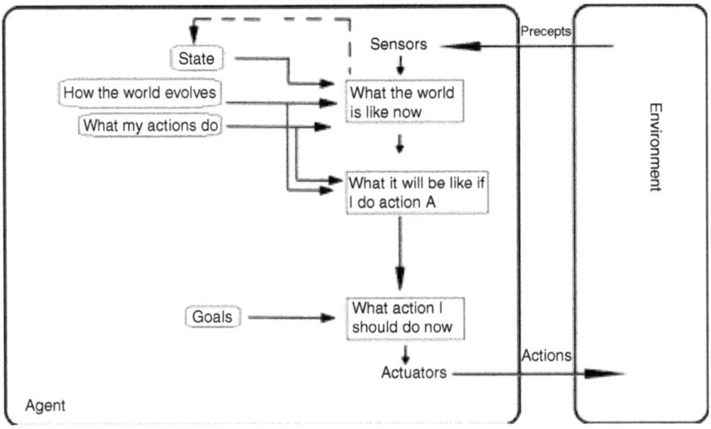

- ✯ Knowledge about current state of the environment.
- ✯ It needs some sort of goal information to choose the action.

  e.g: road function

4. Utility Based Agents:

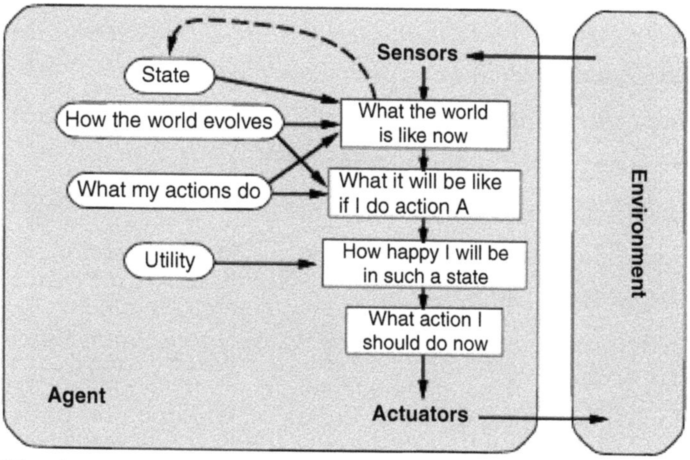

- ☆ Utility functions.
- ☆ Conflicting goals.
- ☆ Several goals.

5. Learning Agents

- ☆ 4 Conceptual Elements:
  - Learning element.
  - Performance element.
  - Critic.
  - Problem generator.

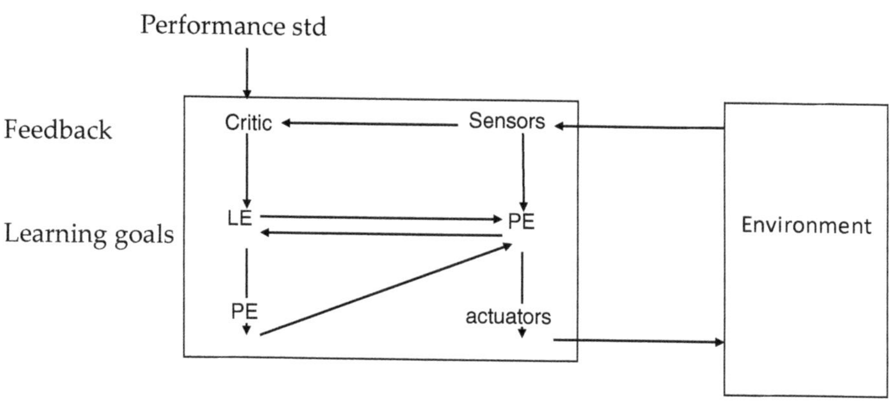

1.11. Problem Solving

PSA

- ☆ Goal based agent who find sequence of actions to achive its goal where as no single action will do.

* Algorithm to solve there problems uniformed search algorithms.

Problem Solving Agents

* Understand the define of a problem.
* Search for finding solutions.
* Maximize the performance measures.

Steps to Maximize the Performance Measures

* Goal formulation (based on current situation, the actions are selected).
* Problem formulation.
* Search (many option of unknown values can examine different possible seq.of actions).
* Solution(problem as input of returns a solution in the form of an action sequence).
* Execution(if there is a solution: it recommends to carried out the actions).

  Once the solution has been executed , the agent will formation a new goal.

Agents in Different Environment

* Static
* Observable
* Discrete
* Deterministic

} Open loop system (ignoring the percepts breaks the loop between agents of the environment)

Search

Formulate a problem as a state space search by showing the legal problem states, the legal operators, and the initial and goal states.

* A state is defined by the specification of the values of all attributes of interest in the world.
* An operator changes one state into the other; it has a precondition which is the value of certain attributes prior to the application of the operator, and a set of effects, which are the attributes altered by the operator.
* The initial state is where you start.
* The goal state is the partial description of the solution.

State Space Search Notations

The set of notations involved in the state space search is :

1. An initial state is the description of the starting configuration of the agent.
2. An action or an operator takes the agent from one state to another state which is called a successor state. A state can have a number of successor states.

3. A plan is a sequence of actions. The cost of a plan is referred to as the path cost.

Factors to Measure the Search Algorithms

1. Complete
2. Optimal
3. What is the search cost associated with the time and memory required to find a solution?
   a. Time complexity
   b. Space complexity

The different search strategies that we will consider include the following:

1. Blind Search strategies or Uninformed search
   a. Depth first search
   b. Breadth first search
   c. Iterative deepening search
   d. Iterative broadening search
2. Informed Search
3. Constraint Satisfaction Search

Blind Search

Blind search or uninformed search that does not use any extra information about the problem domain. The two common methods of blind search are:

* BFS or Breadth First Search
* DFS or Depth First Search

Search Tree is helpful for the searching the goal node.

The terminologies involved in the search tree is:

Root Node: The node from which the search starts.

Leaf Node: A node in the search tree having no children.

Ancestor/Descendant: X is an ancestor of Y is either X is Y's parent or X is an ancestor of the parent of Y. If S is an ancestor of Y, Y is said to be a descendant of X.

Breadth First Search

Let *Queue* be a list containing the initial state

Loop if *Queue* is empty return failure Node remove-first (fringe)

if Node is a goal:

then return the path from initial state to Node

else generate all successors of Node, and

(merge the newly generated nodes into *fringe*)

add generated nodes to the back of *Queue*

End Loop

Advantages of Breadth First Search

Finds the path of minimal length to the goal.

Disadvantages of Breadth First Search

Requires the generation and storage of a tree whose size is exponential the depth of the shallowest goal node.

Depth First Search

Let *Queue* be a list containing the initial state

Loop if *Queue* is empty return failure

Node<-remove-first (*fringe*)

if Node is a goal

then return the path from initial state to Node

else generate all successors of Node,

and merge the newly generated nodes into *Queue*

add generated nodes to the front of *Queue*

End Loop

Depth Limited Search

Let fringe be a list containing the initial state

Loop if fringe is empty return failure

Node<-remove-first (fringe)

if Node is a goal

then return the path from initial state to Node

else if depth of Node = limit return cutoff

else add generated nodes to the front of fringe

End Loop

Depth-First Iterative Deepening (DFID)

until solution found do

DFS with depth cutoff c

c = c+1

A* algorithm

f(n) = g(n) + h(n)

where *g(n)* = sum of edge costs from start to *n*

h(n) = estimate of lowest cost path from

f(n) = actual distan

h(n) is said to be admissible if it underestimates the cost reached from *n*.

If *C*(n)* is the cost of the cheapest

h(n) <= *C*(n)*. we can prove that if *h(n)* is admissible, then the search will find an optimal solution.

A* Algorithm

OPEN = nodes on frontier.

CLOSED = expanded nodes.

OPEN={<S,NIL>} while OPEN is not empty place *<n,p>*

REMOVE FROM OPEN NODE<n.p>with minimum f(n)

if *n* is a goal node,

return success (path p)

for each edge connecting n & M with cost c

if *<m, q>* is on CLOSED and *{p|e}* is cheaper than *q*

then remove *n* from CLOSED,

put *<m,{p|e}>* on OPEN

else if *<m,q>* is on OPEN and *{p|e}* is cheaper than *q*

then replace *q* with *{p|e}*

else if *m* is not on OPEN

then put *<m,{p|e}>* on OPEN

return failure

1.12. Constraint Satisfaction

The general problem is to find a solution that satisfies a set of constraints. The heuristics which are used to decide what node to expand next and not to estimate the distance to the goal. Examples of this technique are design problem, labeling graphs robot path planning and crypt arithmetic puzzles.

In constraint satisfaction problems a set of constraints are available. This is the search space. Initial State is the set of constraints given originally in the problem description. A goal state is any state that has been constrained enough.

Constraint satisfaction is a two-step process:

1. First constraints are discovered and propagated throughout the system.
2. Then if there is not a solution search begins, a guess is made and added to this constraint. Propagation then occurs with this new constraint.

Algorithm

1. Propogate available constraints:
   - ✓ Open all objects that must be assigned values in a complete solution.
   - ✓ Repeat until inconsistency or all objects are assigned valid values:

Select an object and strengthen as much as possible the set of constraints that apply to object.

If set of constraints different from previous set then open all objects that share any of these constraints. Remove selected object.

2. If union of constraints discovered above defines a solution return solution.
3. If union of constraints discovered above defines a contradiction return failure.
4. Make a guess in order to proceed. Repeat until a solution is found or all possible solutions exhausted:
   - ✓ Select an object with a no assigned value and try to strengthen its constraints.
   - ✓ Recursively invoke constraint satisfaction with the current set of constraints plus the selected strengthening constraint.

Crypt arithmetic puzzles are examples of constraint satisfaction problems in which the goal to discover some problem state that satisfies a given set of constraints. Some problems of crypt arithmetic are shown below:

```
   S E N D      D O N A L D        C R O S S
+  M O R E   +  G E R A L D    +   R O A D S
-----------  --------------    -------------
 M O N E Y      R O B E R T      D A N G E R
```

Here each decimal digit is to be assigned to each of the letters in such a way that the answer to the problem is correct. If the same letter occurs more than once it must be assigned the same digit each time. No two different letters may be assigned the same digit.

The puzzle SEND + MORE = MONEY, after solving, will appear like this

```
    S E N D
    9 5 6 7
 +  M O R E
    1 0 8 5
 -----------
  M O N E Y
  1 0 6 5 2
```

State production and heuristics for crypt arithmetic problem.

Ans.

The heuristics and production rules are specific to the following example:

```
    S E N D
    M O R E
  -----------
  M O N E Y
```

1.12.1. Constraint satisfaction problems or CSPs are mathematical problems where one must find states or objects that satisfy a number of *constraints* or criteria. A constraint is a restriction of the feasible solutions in an optimization problem.

A **Constraint Satisfaction Problem** (CSP) is characterized by:

a *set of variables* {x1, x2, .., xn}, for each variable xi a *domain* Di with the possible values for that variable, and a set of *constraints, i.e.* relations, that are assumed to hold between the values of the variables.

[These relations can be given intentionally, *i.e.* as a formula, or extensionally, *i.e.* as a set, or procedurally, *i.e.* with an appropriate generating or recognising function.] We will only consider constraints involving one or two variables.

The constraint satisfaction problem is to find, for each i from 1 to n, a value in Di for xi so that all constraints are satisfied.

A CSP can easily be stated as a sentence in first order logic, of the form:

(exist x1)..(exist xn) (D1(x1) & .. Dn(xn) => C1..Cm)

Generic goal and successor functions:

Generic heuristics (no domain specific expertise).

- ★ An assignment is *complete* when every variable is mentioned.
- ★ A *solution* to a CSP is a complete assignment that satisfies all constraints.
- ★ Some CSPs require a solution that maximizes an *objective function.*

Variables: *WA, NT, Q, NSW, V, SA, T*

- ★ Domains: *Di={red,green,blue}*
- ★ Constraints:adjacent regions must have different colors.

  e.g., WA NT
- ★ Solutions are assignments satisfying all constraints,

  e.g., {WA=red,NT=green,Q=red,NSW=green,V=red,SA=blue,T=green}
- ★ Constraint graph: • nodes are variables • arcs are binary constraints

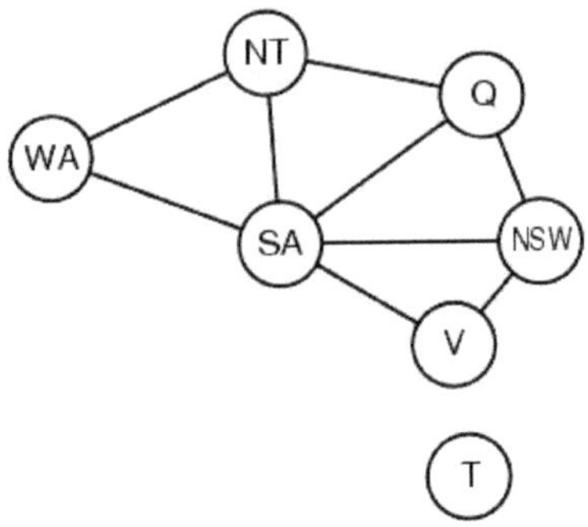

Graph can be used to simplify search

e.g., Tasmania is an independent subproblem

Types of Variables

Discrete Variables

– Finite domains; size *d O(dn)* complete assignments.

e.g., Boolean CSPs: Boolean satisfiability (NP-complete).

– Infinite domains (integers, strings, etc.)

e.g., job scheduling, variables are start/end days for each job

Need a constraint language *e.g., StartJob1 +5 ≤ StartJob3.*

Infinitely many solutions

Linear constraints: solvable

Nonlinear: no general algorithm

Continuous Variables

– *e.g.,* building an airline schedule or class schedule.

– Linear constraints solvable in polynomial time by LP methods.

– Unary constraints involve a single variable.

e.g., SA green

– Binary constraints involve pairs of variables.

e.g., SA WA

CSP as SEARCH

- ✯ A CSP can easily be expressed as a standard search problem.
- ✯ Incremental formulation
  - *Initial State*: the empty assignment {}.
  - *Successor function*: Assign a value to any unassigned variable provided that it does not violate a constraint.
  - *Goal test*: the current assignment is complete.
  - *Path cost*: constant cost for every step.

CSP by Backtracking

- ✯ Similar to Depth-first search
- ✯ Chooses values for one variable at a time and backtracks when a variable has no legal values left to assign.
- ✯ Uninformed algorithm **function** BACKTRACKING-SEARCH(*csp*) **return** a solution or

Failure

return RECURSIVE-BACKTRACKING(*{}* , *csp*)

function RECURSIVE-BACKTRACKING(*assignment, csp*) **return** a solution or failure

if *assignment* is complete **then return** *assignment*

var SELECT-UNASSIGNED-VARIABLE(VARIABLES[*csp*],*assignment*,*csp*)

for each *value* **in** ORDER-DOMAIN-VALUES(*var, assignment, csp*) **do**

if *value* is consistent with *assignment* according to CONSTRAINTS[*csp*] **then** add *{var=value}*

to assignment

result RRECURSIVE-BACTRACKING(*assignment, csp*)

if *result failure* **then return** *result*

remove *{var=value}* from *assignment*

return *failure*

2

Representation of Knowledge

2.1. Game Playing

Game Playing is one of the oldest sub-fields in AI. Game playing involves abstract and pure form of competition that seems to require intelligence. It is easy to represent the states and actions. To implement the game playing very little world knowledge is required.

The most common used AI technique in game is search. Game playing research has contributed ideas on how to make the best use of time to reach good decisions.

Game playing is a search problem defined by:

- ⋆ Initial state of the game
- ⋆ Operators defining legal moves
- ⋆ Successor function
- ⋆ Terminal test defining end of game states
- ⋆ Goal test
- ⋆ Path cost/utility/payoff function

More popular games are too complex to solve, requiring the program to take its best guess, for example in chess, the search tree has 1040 nodes (with branching factor of 35). It is the opponent because of whom uncertainty arises.

Characteristics of Game Playing

1. There are always an "unpredictable" opponent:
    - The opponent introduces uncertainty.
    - The opponent also wants to win.

The solution for this problem is a strategy, which specifies a move for every possible opponent reply.

2. Time limits:

- ✯ Optimal decisions
- ✯ α-β pruning
- ✯ Imperfect, real-time decisions

Games vs. Search Problems

- ✯ "Unpredictable" opponent → specifying a move for every possible opponent reply
- ✯ Time limits → unlikely to find goal, must approximate

Game tree (2-player, deterministic, turns)

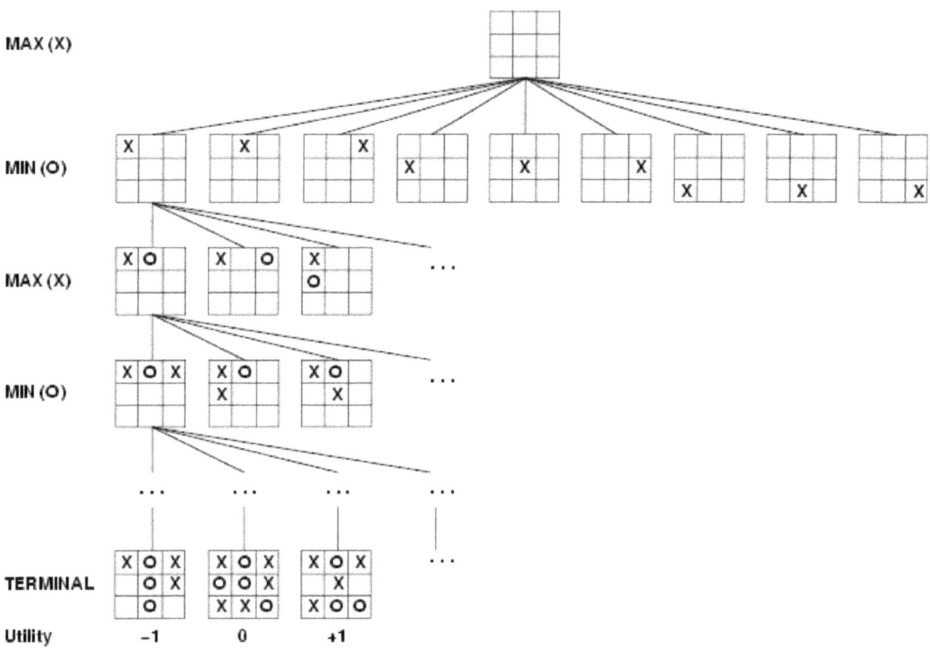

Optimal Strategy

- ✯ Perfect play for deterministic games
- ✯ **Minimax Value** for a node n

MINIMAX-VALUE(n)=

UTILITY(n)	If n is a terminal
$\max_{s \in successors(n)}$ MINIMAX-VALUE(s)	If n is a max node
$\min_{s \in successors(n)}$ MINIMAX-VALUE(s)	If n is a min node

- ★ This definition is used recursively
- ★ Idea: minimax value is the best achievable payoff against best play

Minimax Example

- ★ Perfect play for deterministic games.
- ★ **Minimax Decision** at root: choose the action a that lead to a maximal minimax value.
- ★ MAX is guaranteed for a utility which is at least the minimax value – if he plays rationally.

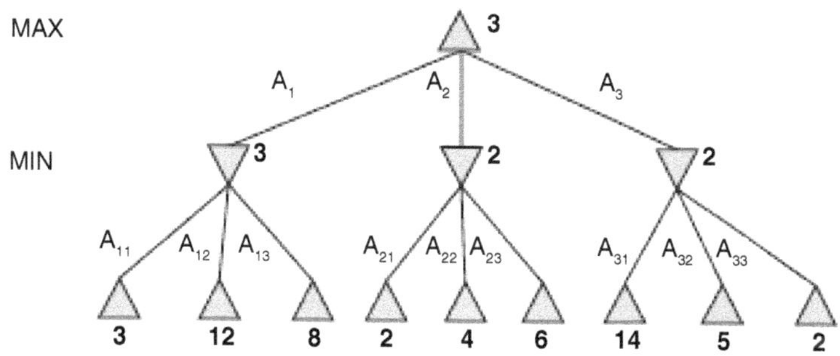

2.1.1. Optimal Decisions in Games

- ★ Consider games with two players (MAX, MIN)
- ★ Initial State

  Board position and identifies the player to move
- ★ Successor Function

  Returns a list of (move, state) pairs; each a legal move and resulting state
- ★ Terminal Test

  Determines if the game is over (at terminal states)
- ★ Utility Function

  Objective function, payoff function, a numeric value for the terminal states (+1, -1) or (+192, -192)

2.2. Knowledge Representation

- ★ ***Logic*** provides the formal structure and rules of inference.
- ★ ***Ontology*** defines the kinds of things that exist in the application domain.
- ★ ***Computation*** supports the applications that distinguish knowledge representation from pure philosophy.

Knowledge Representation Languages

High level representation formalisms, that can in principle be implemented using programming languages. should **support** ***inference***. ***inference*** - any way to get new expressions from old.

Requirements

- ⋆ **Representational Adequacy:** should allow to represent the knowledge we need
- ⋆ **Inferential Adequacy:** ability to infer new knowledge from a basic set of facts
- ⋆ **Inferential Efficiency:** ability to incorporate additional information into the knowledge structure which can be used to focus the attention of the inference mechanism in the most promising direction.
- ⋆ **Clear Syntax and Semantics**: what the allowable expressions are and what they mean
- ⋆ **Naturalness:** natural and easy to use

2.2.1. What is a Logic?

- ⋆ A language with concrete rules
  - No ambiguity in representation (may be other errors!)
  - Allows unambiguous communication and processing
  - Very unlike natural languages, *e.g.* English
- ⋆ Many ways to translate between languages
  - A statement can be represented in different logics
  - And perhaps differently in same logic
- ⋆ **Expressiveness** of a logic
  - How much can we say in this language?
- ⋆ Not to be confused with logical reasoning

Logics are languages, reasoning is a process (may **use** logic)

Syntax and Semantics

- ⋆ Syntax
  - Rules for constructing legal sentences in the logic
  - Which symbols we can use (English: letters, punctuation)
  - How we are allowed to combine symbols
- ⋆ Semantics
  - How we interpret (read) sentences in the logic
  - Assigns a meaning to each sentence
- ⋆ Example: "All lecturers are seven foot tall"
  - A valid sentence (syntax)
  - And we can understand the meaning (semantics)
  - This sentence happens to be false (there is a counterexample)

Propositional Logic

- ☆ Syntax
  - Propositions, *e.g.* "it is wet"
  - Connectives: and, or, not, implies, iff (equivalent)
  - Brackets, T (true) and F (false)
- ☆ Semantics (Classical AKA Boolean)
  - Define how connectives affect truth

    "P and Q" is true if and only if P is true and Q is true
  - Use **truth tables** to work out the truth of statements

2.3. Predicate Logic

- ☆ Propositional logic combines atoms
  - An atom contains no propositional connectives
  - Have no structure (today_is_wet, john_likes_apples)
- ☆ **Predicates** allow us to talk about objects
  - Properties: is_wet(today)
  - Relations: likes(john, apples)
  - True or false
- ☆ In predicate logic each atom is a predicate
  - *e.g.*, first order logic, higher-order logic

Prolog Programming

- ☆ Declaring facts about objects and their relationships -> likes (john,mary)
- ☆ Defining rules about objects and relationships
- ☆ Asking Questions about objects

Example

```
sister-of(X,Y) :- female(X),
                  parents(X,M,F),
                  parent(Y,M,F)
```

Ontological Commitment

What exists in the world – TRUTH

PL -> Facts hold or donot hold

FOL -> Objects with relations between them that hold or do not hold.

Elements

- ☆ Constant symbols -> stand for objects.
- ☆ Predicate symbols -> stand for relations.

- Function symbols -> stand for functions.

A FOL model is basically equivalent to relational database instance.

First Order Logic

- Representation language of knowledge.
- Powerful than propositional language.
- Boolean logic.
- Expressive, declarative and compositional language.
- Propositional logic assumes that the world contains facts.

First-order logic (like natural language) assumes the world contains.

- Objects: people, houses, numbers, colors, baseball games, wars.
- Relations: red, round, prime, brother of, bigger than, part of, comes between.
- Functions: father of, best friend, one more than, plus, ...

Examples

- Objects: Students, lectures, companies, cars ...
- Relations: Brother-of, bigger-than, outside, part-of, has-color, occurs-after, owns, visits, precedes, ...
- Properties: blue, oval, even, large, ...
- Functions: father-of, best-friend, second-half, one-more-than ...

Knowledge Representation Schemes

There are four types of knowledge representation:

Relational, Inheritable, Inferential, and Declarative/Procedural.

Relational Knowledge

- Provides a framework to compare two objects based on equivalent attributes.
- Any instance in which two different objects are compared is a relational type of knowledge.

Inheritable Knowledge

- Is obtained from associated objects.
- It prescribes a structure in which new objects are created which may inherit all or a subset of attributes from existing objects.

Inferential Knowledge

- Is inferred from objects through relations among objects.
- *e.g.*, a word alone is a simple syntax, but with the help of other words

in phrase the reader may infer more from a word; this inference within linguistic is called semantics.

Declarative Knowledge

- ✯ A statement in which knowledge is specified, but the use to which that knowledge is to be put is not given.
- ✯ *e.g.*, laws, people's name; these are facts which can stand alone, not dependent on other knowledge.

Procedural Knowledge

- ✯ A representation in which the control information, to use the knowledge, is embedded in the knowledge itself.
- ✯ *e.g.*, computer programs, directions, and recipes; these indicate specific use or implementation;

Relational Knowledge

This knowledge associates elements of one domain with another domain.

- ✯ Relational knowledge is made up of objects consisting of attributes and their corresponding associated values.
- ✯ The results of this knowledge type is a mapping of elements among different domains.

The table below shows a simple way to store facts.

- ✯ The facts about a set of objects are put systematically in columns.
- ✯ This representation provides little opportCHAPTERy for inference.

Table -	Height	Weight	Bats - Throws
Aaron	6-0	180	Right - Right
Mays	5-10	170	Right - Right
Ruth	6-2	215	Left - Left
Williams	6-3	205	Left - Right

Simple Relational Knowledge Player

2.3.1. Approaches to Knowledge Representation

- ✯ Declarative vs procedural knowledge
- ✯ **Declarative representations:**
  - Logic
  - Semantic networks and Frames
  - Production Rules

First Order Predicate Logic

- ✯ FOPL has a well defined syntax and semantics.
- ✯ It is concerned with truth preserving inference.
- ✯ Problems : time, beliefs and uncertainty are difficult to represent.

Semantic Nets and Frames:

- ✯ Represent factual knowledge about classes of objects and their properties.
- ✯ Not formal systems.
- ✯ Basic inference mechanism: inheritance of properties.

Problems: Quantifiers, Representing Disjunction and Negation

Production Rules

Production Systems:

A set of if-then rules - typically state that if certain conditions hold, then some action should be taken.

If -then Relation:

If high_temperature then prescribe aspirin.

Production systems use a working memory - represents the facts (as semantic nets or frames) that are currently believed to hold.

Semantic Nets

A semantic net is represented as a **graph**, where the nodes in the graph represent concepts, and the arcs represent binary relationships between concepts.

- ✯ **Nodes** represent objects, attributes and values.
- ✯ **Links** represent attributes and relationships between nodes.
- ✯ **Labels** attached to links: the name of the corresponding attribute or relation

2.3.2. Ontology Engineering

- ✯ Ontology engineering is a set of tasks related to the development of ontologies for a particular domain.
- ✯ Defining terms in the domain and relations among them:
  - Defining concepts in the domain (classes).
  - Arranging the concepts in a hierarchy (subclass-superclass hierarchy).
  - Defining which attributes and properties (slots) classes can have and constraints on their values.
  - Defining individuals and filling in property values.
  - Ontology Engineering analogous to OO Database Design?

Differences with Special-Purpose Ontologies

A general-purpose ontology should be applicable in more or less any special-purpose domain.

- ✯ Add domain-specific axioms.

In any sufficiently demanding domain different areas of knowledge need to be unified.

- ☆ Reasoning and problem solving could involve several areas simultaneously.

  What do we need to express?

  - Categories, Measures, Composite objects, Time, Space, Change, Events, Processes, Physical Objects, Substances, Mental Objects, Beliefs.

2.3.3. Categories and Objects

- ☆ KR requires the organization of objects into categories:
  - Interaction at the level of the object
  - Reasoning at the level of categories
- ☆ Categories play a role in predictions about objects:
  - Based on perceived properties
- ☆ Categories can be represented in two ways by first-order logid (FOL):
  - Predicates: apple(x).
  - Reification of categories into objects: apples.
- ☆ Category = set of its members.

FOL and Categories

- ☆ An object is a member of a category:
  - Member of (BB12,Basketballs).
- ☆ A category is a subclass of another category:
  - Subset of (Basketballs,Balls).
- ☆ All members of a category have some properties:
  - $\forall$ x (MemberOf(x,Basketballs) $\Rightarrow$ Round(x))
- ☆ All members of a category can be recognized by some properties:
  - $\forall$ x (Orange(x) $\wedge$ Round(x) $\wedge$ Diameter(x)=9.5in $\wedge$ MemberOf(x,Balls) $\Rightarrow$ MemberOf(x,BasketBalls))
- ☆ A category as a whole has some properties:
  - MemberOf(Dogs,DomesticatedSpecies)

2.4. Resolution

- ☆ Convert everything to CNF.
- ☆ Resolve, with unification.
- ☆ If resolution is successful, proof succeeds.
- ☆ If there was a variable in the item to prove, return variable's value from unification bindings.

- ★ Resolution allows a complete inference mechanism (search-based) using only one rule of inference
- ★ Resolution rule:
  - Given: $P_1 \vee P_2 \vee P_{3\ \ldots} \vee P_{n,}$ and $\vee P_1 \vee Q_{1\ \ldots}\ Q_m$
  - Conclude: $P_2 \vee P_{3\ \ldots} \vee P_n \vee Q_{1\ \ldots} \vee Q_m$

  Complementary literals P_1 and $\vee P_1$ "cancel out"
- ★ To prove a proposition F by resolution:
  - Start with ∨F
  - Resolve with a rule from the knowledge base (that contains F)
  - Repeat until all propositions have been eliminated
  - If this can be done, a contradiction has been derived and the original proposition F must be true.

Propositional Resolution Example

- ★ Rules

  Cold and precipitation -> snow

  ¬cold ∨¬precipitation ∨ snow

  January -> cold

  ¬January ∨ cold

  Clouds -> precipitation

  ¬clouds ∨ precipitation
- ★ Facts

  – January, clouds
- ★ Prove

  – snow

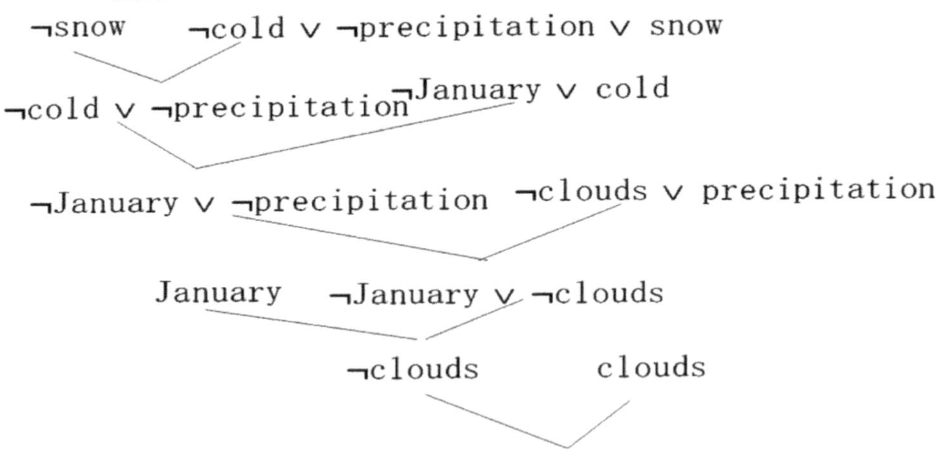

Resolution Theorem Proving (FOL)

- Convert everything to CNF.
- Resolve, with unification.
  - Save bindings as you go!
- If resolution is successful, proof succeeds.
- If there was a variable in the item to prove, return variable's value from unification bindings.

Converting to CNF

1. Replace implication $(A \Rightarrow B)$ by $\neg A \vee B$
2. Move $\neg$ "inwards":
   - $\neg\forall$ x P(x) is equivalent to $\exists$ x $\neg$P(x) & vice versa
3. Standardize variables:
   - $\forall$ x P(x) $\forall\neg$ x Q(x) becomes $\forall$ x P(x) $\vee$ $\forall$y Q(y)
4. Skolemize:
   - $\exists$ x P(x) becomes P(A)
5. Drop universal quantifiers:
   - Since all quantifiers are now $\forall$, we don't need them
6. Distributive Law

Convert to FOPL, then CNF

- John likes all kinds of food
- Apples are food.
- Chicken is food.
- Anything that anyone eats and isn't killed by is food.
- Bill eats peanuts and is still alive.
- Sue eats everything Bill eats.

Prove Using Resolution

- John likes peanuts.
- Sue eats peanuts.
- Sue eats apples.
- What does Sue eat?
  - Translate to Sue eats X
  - Result is a valid binding for X in the proof
- Resolution is complete. If you don't want to take this on faith.
- Strategies for efficient resolution include:

- CHAPTER preference. If a clause has only one literal, use it first.
- Set of support. Identify "useful" rules and ignore the rest.
- Input resolution. Intermediately generated sentences can only be combined with original inputs or original rules.
- Subsumption. Prune unnecessary facts from the database.

Use of Predicate Calculus

First Order Logic

- ✯ More expressive logic than propositional
  - Used in this course (Lecture 6 on representation in FOL)
- ✯ **Constants** are objects: john, apples
- ✯ **Predicates** are properties and relations:
  - likes(john, apples)
- ✯ **Functions** transform objects:
  - likes(john, fruit_of(apple_tree))
- ✯ **Variables** represent any object: likes(X, apples)
- ✯ **Quantifiers** qualify values of variables:
  - True for all objects (Universal): ∀X. likes(X, apples)
  - Exists at least one object (Existential): ∃X. likes(X, apples)
- ✯ Propositional logic assumes that the world contains facts.
- ✯ FOL assumes that world contains.
  - Objects -> Peoples , houses , colors , games , no.s .
  - Relations -> Red , round , Prime , Part of , Comes from , between .
  - Functions -> Father of , Best friend , one more than , plus .

2.5 Well Defined Problem & Solutions

A problem can be defined by 4 components

- ✯ <u>Initial state</u>
- ✯ <u>Successor function</u>: Description of the possible actions:
  - State space
  - Path
  - <u>Goal test</u>: Check whether any one of the goal stae is reached (or) not.

    Chess→ Goal to reach a state →"checkmate".
  - <u>Path cost</u> : Function that assigns a numeric cost to the path

    C(X,a,Y), c→cost

    X,Y→States

    a→actions

steps cost are non-negative.

→A Solution to a problem is a path from initial state to a goal state.

→Solution quality is measured by path cost function:

Total cost=path cost+search cost

Formulating problems:

e.g: Route Finding Problems

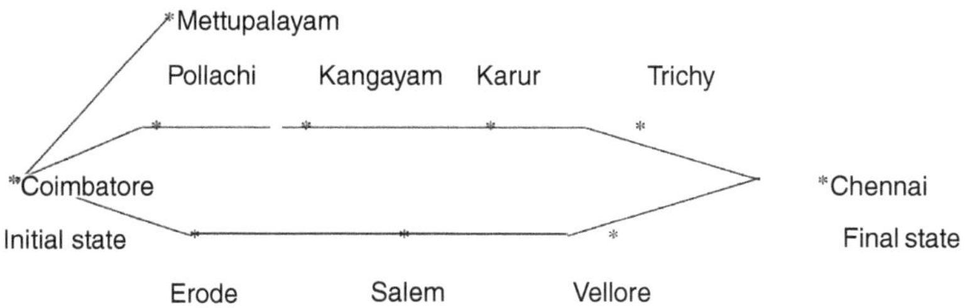

Successor Function

{

< Go (Pollachi) , In (Pollachi)>

Go (Erode),In (Erode)>

Go(palladam),In (palladam)>

Go (Mettupalayam),In(mettupalayam)>

}

<u>Goal Test</u>→ Singleton sct{in(chennai)}

<u>Path cost</u>→{(in coimbatore)}

{go(erode),}=100[Killometers]

{in(erode}

- ★ Solution to a problem is a path from initial state to goal state.
- ★ Solution performance is measured by path cost function.
- ★ An optimal solution has lowest path cost among all solutions.

Abstraction

- ★ Process of Removing Detail from a representation called abstraction.

e.g: Route Finding Problems

- Travelling Companions
- Scenery out of window
- Condition of road
- Weather

} irrelevant to find way to Chennai

Example Problems

- ★ TOY Problems
- ★ Real world problems

Toy Problems

- ★ Vaccum World:

→goal is to clean the dirty in square by using the vaccum cleaner.

States:

2X2^2=8 possible states

(might or might not contain dirts).

Goal test → checks whether all the squares are clean

Path cost → steps cost 1 , path cost is the no of steps in the path .

→ N locations with 2^n states (Toy problem has discrete locations).

8-puzzle

3*3 board with eight numbered tiles of a blank space

→Objective is to reach the desired goal state.

7	2	4
5		6
8	3	1

Start state

	1	2
3	4	5
6	7	8

Goal state

States → specifies the location of each of 8 titles of the blank space.

Initial states → any state can be designated as I.S.

Successor Function → General the legal states , four actions (Blank moves, left , right , up or down)

Goal test → checks whether the state matches the goal configuration.

Path cost → Each steps cost 1 , the path cost is the no of steps in the path.

– Actions are abstracted

8 – puzzle problem has 9!/2=1,81,440 reaching stages and easily solved.

Belongs to sliding block puzzles.

8-Queens Problem

- ★ Goal of the problem is to place 8 queens on a chessboard.

★ No queen attack each other in row wise , column wise or diagonal.

Q							
		Q					
				Q			
						Q	
	Q						
			Q				
					Q		
							Q

2 kinds of formulation,

1. Incremental
2. Complete state formulation

1) Incremental formulation → Starts with an empty state for the 8-queens problem.

Complete state → starts with all the 8-queens on the board and move them around.

States → Any arrangements 0 to 8 queens on the board.

I.S → No queens on the board.

Successor Functions

→ add a queen to an empty square.

Goal Test

→ 8 queen are on the board but none of them should attack each other.

2) Problem Components for Complete State Formulation States

Arrangements of n queens ($0 \le n \le 8$)

One per column in the left most n columns , with no queen attacking another .

Successor Function

Add a queen to any square in the leftmost empty column , it is not attacked any other queen.

→ reduces the state space from 3 *(10^14) for the earlier formulation.

Real World Problems

It is a problem where, people actually care for solutions.

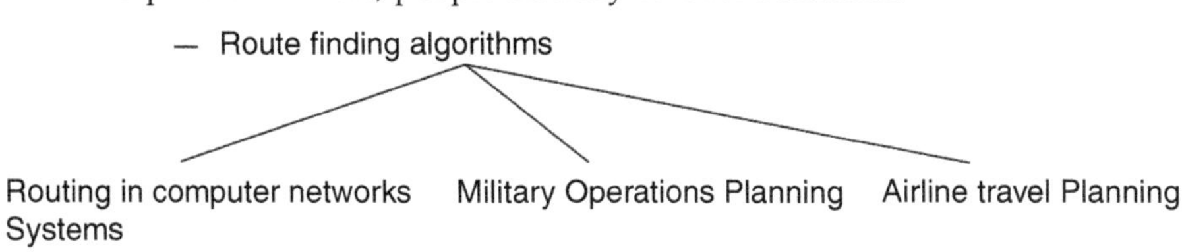

a) Airline Travel Problem

States: Each state is represented by a location and the current time.

Initial state: It is specified by the problem.

Successor Function

Returns the states resulting from taking any scheduled flight, leaving later than the current time ⊕ the transmit time within airport.

Goal Test → checks whether the destination is reached in prescribed time.

Path cost : Depends on Monetory cost Waiting time, Flight time, seat quality, time of day, type of air plane.

Searching for solutions

Problems can be solved by searching through the state space.

→ A search tree is generated by the initial state and the successor function that together define the state space.

→ When same state can be reached from multiple paths, we have a search graph rather than a search tree.

Search Tree Algorithm

Example: Route finding

Search tree → Tree which is drawn to find the route from Coimbatore to Chennai.

Search node → Root of the search tree is a search node corresponding to initial state.

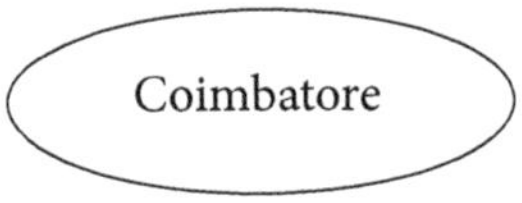

Start the sequence with the initial state of check whether it is goal state or not.

*If it is a goal state, return success.

*otherwise perform the following sequence of steps,

→ Search strategy.

→ If the goal state is not reachable, next step is expanding the current state.

→ New states are generated and expanded.

→ If we choose pollachi first, we check whether it is a goal state. It is not, then return to:

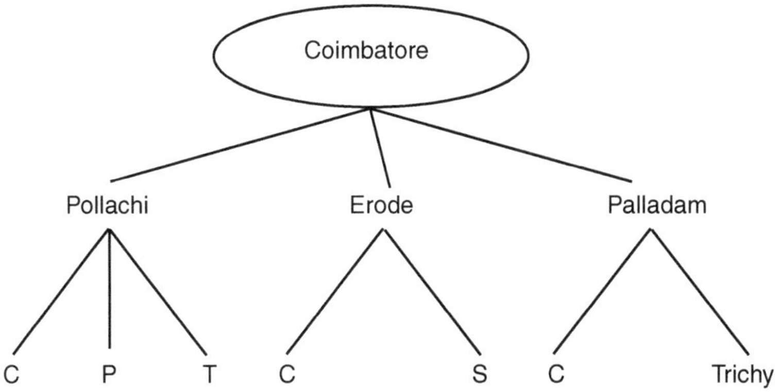

Data structure representation for search trees, Nodes in the search tree defined using 5 components,

- ✭ START
- ✭ PARENT-NODE
- ✭ ACTION
- ✭ PATH-COST
- ✭ DEPTH

Measuring Problem-Solving Performance

Algorithm's performance is evaluated in 4 ways:

*completeness

*optimality

*Time complexity

*Space complexity

Three quantities of time complexity and space complexity:

1. Branching factor
2. Depth
3. Maximum length

Effectiveness of Search Algorithm

1. Search cost
2. Total cost.

Backtracking Search

Heuristics are used to determine which variable to assign next "Pick Unassigned Variable".

- ✭ The choice can vary from branch to branch, *e.g.*, under the assignment V1=a we might choose to assign V4 next, while under V1=b we might choose to assign V5 next.

- ★ This "dynamically" chosen variable ordering has a tremendous impact on performance.
  - → Similar to DFS.
  - → Choose values for one variable at a time and backtracks when a variable has no legal values left to assign.
  - → Uninformed algorithm.

Backtracking (BT) Algorithm

```
BT(Level)
If all variables assigned
PRINT Value of each Variable
RETURN or EXIT (RETURN for more solutions)
(EXIT for only one solution)
V := PickUnassignedVariable()
Variable[Level] := V
Assigned[V] := TRUE
for d := each member of Domain(V)
Value[V] := d
OK := TRUE
for each constraint C such that
V is a variable of C
and all other variables of C are assigned.
if C is not satisfied by the current set of assignments
OK := FALSE
if(OK)
BT(Level+1)
return
```

Example

N---Queens

- ★ Place N Queens on an N X N chess board so that no Queen can attack any other Queen.

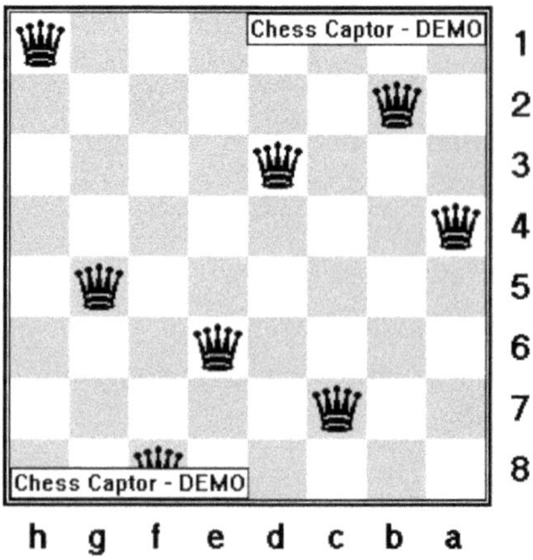

Problem Formulation

- ★ N variables (N queens)
- ★ N2 values for each variable representing the positions on the chessboard.

- ✯ Value i is i'th cell counting from the top left as 1, going left to right, top to bottom.

Q1 = 1, Q2 = 15, Q3 = 21, Q4 = 32, Q5 = 34, Q6 = 44, Q7 = 54, Q8 = 59

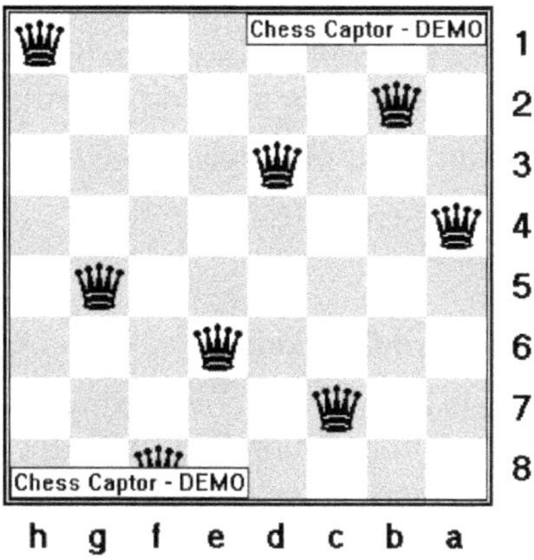

This representation has (N2)N states (different possible assignments in the search space)

- ✯ For 8-Queens: 64^8 = 281,474,976,710,656:
  - Is there a better way to represent the Nqueens problem?
- ✯ We know we cannot place two queens in a single row.

We can exploit this fact in the choice of the CSP representation.

Better Modeling

- ✯ N variables Qi, one per row.
- ✯ Value of Qi is the column the Queen in row i is placed; possible values {1, …, N}.
- ✯ This representation has NN states:
- ✯ For 8-Queens: 8^8 = 16,777,216
- ✯ The choice of a representation can make the problem solvable or unsolvable!

2.6. Constraint Propagation

- ✯ Propagation has to be applied during search.
- ✯ Potentially at every node of the search tree:
  - If propagation is slow, this can slow the search down to the point where we are better off not to do any propagation! there is always a trade off between searching fewer nodes in the search, and having a higher nodes/second processing rate.
  - Propagation has to be applied during the search; potentially at every node of the search tree.

- Propagation itself is an inference step that needs some resources (in particular time)
- If propagation is slow, this can slow the search down to the point where using propagation makes finding a solution take longer!
- There is always a tradeoff between searching fewer nodes in the search, and having a higher nodes/ second processing rate.

We will look at two main types of propagation:

★ Forward Checking & Generalized Arc

★ Consistency

Constraint Propagation: Forward Checking

★ Forward checking is an extension of backtracking search that employs a "modest" amount of propagation (look ahead).

★ When a variable is instantiated we check all constraints that have only one uninstantiated variable remaining.

★ For that uninstantiated variable, we check all of its values, pruning those values that violate the constraint.

Local Search

★ Use complete state representation.

★ Allow states with unsatisfied constraints (unlike backtracking).

- Variable selection → Randomly select any conflicted variable.
- Value selection → Minimum conflicts heuristic.

Advantages

*Run time of minimum conflicts is roughly independent of problem size.

Forward checking → prevents assignments that guarantee later failure.

Constraint paragraph → propagating the implications of a constraints on one variable onto other variables.

2.7. Intelligent Backtracking

→ Looking backward.

Example – Map Colouring

★ Color the following map using *red*, *green*, and *blue* such that adjacent regions have different colors.

Constraint Satisfaction Problem

Modeling

- ✯ Variables: *WA, NT, Q, NSW, V, SA, T*
- ✯ Domains: *Di={red, green, blue}*
- ✯ Constraints: adjacent regions must have different colors.

  e.g. WA ≠ NT

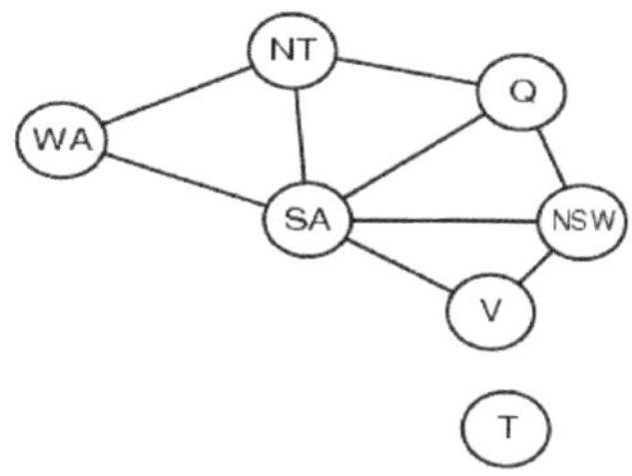

Forward checking idea: keep track of remaining legal values for unassigned variables.

- ✯ Terminate search when any variable has no legal values.

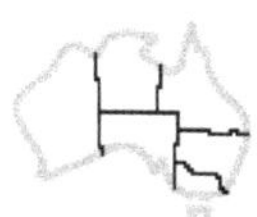

Assign *{WA=red}*

- ✯ Effects on other variables connected by constraints to WA
  - *NT can no longer be red*
  - *SA can no longer be red*

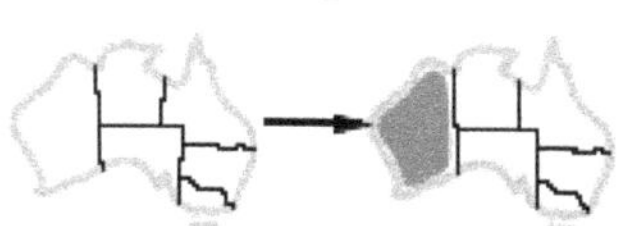

Assign *{Q=green} (Note: Not using MRV)*

- ★ Effects on other variables connected by constraints with Q
    - *NT can no longer be green*
    - *NSW can no longer be green*
    - *SA can no longer be green*
- ★ *MRV heuristic* would automatically select NT or SA next

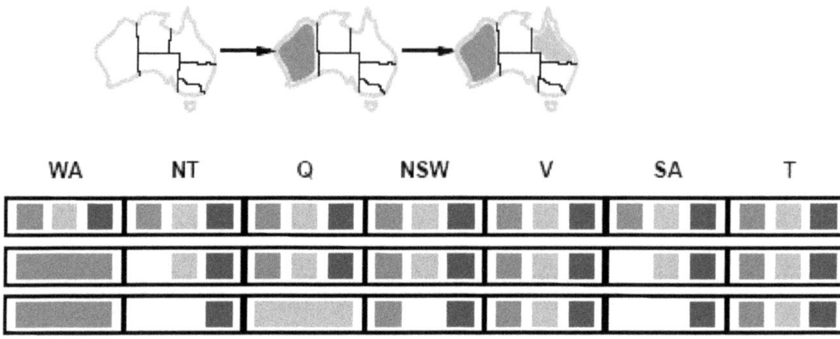

Assign *{V=blue} (not using MRV)*

- ★ Effects on other variables connected by constraints with V
    - *NSW can no longer be blue*
    - *SA is empty*
- ★ FC has detected that partial assignment is *inconsistent* with the constraints and backtracking can occur.

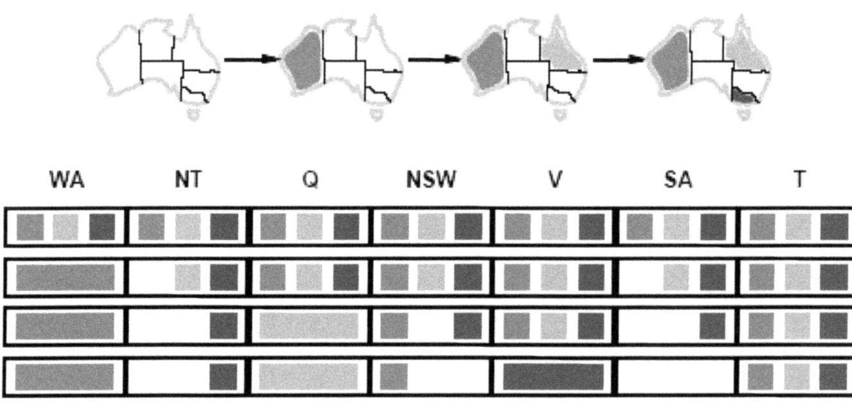

Many Real World Applications of CSP

- ★ Assignment problems – who teaches what class.
- ★ Timetabling problems – exam schedule.
- ★ Transportation scheduling.
- ★ Floor planning

- ✯ Factory scheduling
- ✯ Hardware configuration – a set of compatible components

Why Games?

- ✯ Small defined set of rules.
- ✯ Well defined knowledge set.
- ✯ Easy to evaluate performance.
- ✯ Large search spaces.
  - Too large for exhaustive search.
- ✯ Fame and Fortune.

  e.g. Chess and Deep Blue

Types of Games

	Deterministic	Chance
Perfect Information	Chess, checkers, go, othello	Backgammon, monopoly
Imperfect Information		Bridge, poker, scabble, nuclear war

2.8. Alpha-beta Pruning

Alpha-beta pruning is a search algorithm which seeks to reduce the number of nodes that are evaluated by the minimax algorithm in its search tree. It is a search with adversary algorithm used commonly for machine playing of two-player games (Tic-tac-toe, Chess, Go, etc.). It stops completely evaluating a move when at least one possibility has been found that proves the move to be worse than a previously examined move. Such moves need not be evaluated further.

Definition

Alpha-beta pruning is a sound optimization in that it does not change the score of the result of the algorithm it optimizes.

Why is it called α-β?

- ✯ α is the value of the best (i.e., highest-value) choice found so far for *MAX* at any choice point along the path to the root.
- ✯ If v is worse than α, *MAX* will avoid it → prune that branch
- ✯ β is the value of the best (*i.e.*, lowest-value) choice found so far for MIN at any choice point along the path for to the root.
- ✯ Recognize when a position can never be chosen in minimax no matter what its children are:
  - Max (3, Min(2,x,y) …) is always ≥ 3.
  - Min (2, Max(3,x,y) …) is always ≤ 2.
  - We know this without knowing x and y!

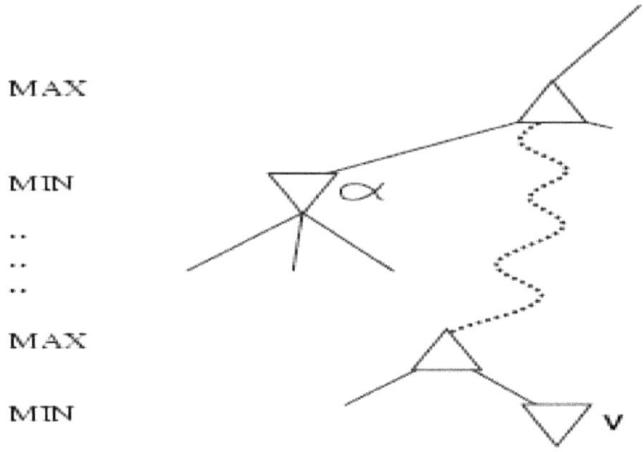

Benefits

- ★ Without pruning, need to examine $O(b^m)$ nodes.
- ★ With pruning, depends on which nodes we consider first.
- ★ If we choose a random successor, need to examine $O(b^{3m/4})$ nodes.
- ★ If we manage to choose the best successor first, need to examine $O(b^{m/2})$ nodes.
- ★ Practical heuristics for choosing next successor to consider get quite close to this.
- ★ Can effectively look twice as deep!

2.9. Stochastic Games

- ★ Games played perfectly:
    - Connect four, noughts & crosses (tic-tac-toe).
    - Best move pre-calculated for each board state.
    - Small number of possible board states.
- ★ Games played well:
    - Chess, draughts (checkers), backgammon.
    - Scrabble, tetris (using ANNs).
- ★ Games played badly:
    - Go, bridge, soccer.

Techniques

- ★ **Transposition table** to store previously expanded states.
- ★ **Forward pruning** to avoid considering all possible moves.
- ★ **Lookup tables** for opening moves and endgames.

The State of the Art for Some Games

- Chess:
  - 1997: IBM Deep Blue defeats Kasparov.
  - ... there is still debate about whether computers are really better.
- Checkers:
  - Computer world champion since 1994.
  - ... there was still debate about whether computers are really better...
  - until 2007: checkers solved optimally by computer.
- Poker:
  - Competitive with top humans in some 2-player games.
  - 3+ player case much less well-understood.

2.10. Structured Represntation of Knowledge

- Representing knowledge using logical formalism, like predicate logic, has several advantages. They can be combined with powerful inference mechanisms like resolution, which makes reasoning with facts easy. But using logical formalism complex structures of the world, objects and their relationships, events, sequences of events, etc. cannot be described easily.
- A good system for the representation of structured knowledge in a particular domain should posses the following four properties:
  - i. Representational Adequacy: The ability to represent all kinds of knowledge that are needed in that domain.
  - ii. Inferential Adequacy: The ability to manipulate the represented structure and infer new structures.
  - iii. Inferential Efficiency: The ability to incorporate additional information into the knowledge structure that will aid the inference mechanisms.
  - iv. Acquisitional Efficiency: The ability to acquire new information easily, either by direct insertion or by program control.

The techniques that have been developed in AI systems to accomplish these objectives fall under two categories:

1. **Declarative Methods**: In these knowledge is represented as static collection of facts which are manipulated by general procedures. Here the facts need to be stored only one and they can be used in any number of ways. Facts can be easily added to declarative systems without changing the general procedures.
2. **Procedural Method**: In these knowledge is represented as procedures. Default reasoning and probabilistic reasoning are examples of procedural methods. In these, heuristic knowledge of "How to do things efficiently"can be easily represented.

In practice most of the knowledge representation employ a combination of both. Most of the knowledge representation structures have been developed to handle programs that handle natural language input.

One of the reasons that knowledge structures are so important is that they provide a way to represent information about commonly occurring patterns of things. Such descriptions are some times called schema.

One Definition of Schema is

- ✭ "Schema refers to an active organization of the past reactions, or of past experience, which must always be supposed to be operating in any well adapted organic response".
- ✭ By using schemas, people as well as programs can exploit the fact that the real world is not random. There are several types of schemas that have proved useful in AI programs. They include:
  - i. Frames: Used to describe a collection of attributes that a given object possesses (*e.g*: description of a chair).
  - ii. Scripts: Used to describe common sequence of event (*e.g*:, a restaurant scene).
  - iii. Stereotypes: Used to described characteristics of people.
  - iv. Rule models: Used to describe common features shared among a set of rules in a production system.
- ✭ Frames and scripts are used very extensively in a variety of AI programs. Before selecting any specific knowledge representation structure, the following issues have to be considered.
  - i. The basic properties of objects, if any, which are common to every problem domain must be identified and handled appropriately.
  - ii. The entire knowledge should be represented as a good set of primitives.
  - iii. Mechanisms must be devised to access relevant parts in a large knowledge base.

3

Knowledge Inference

3.1. Knowledge Representation

The object of a knowledge representation is to express knowledge in a computer tractable form, so that it can be used to enable our AI agents to perform well.

A knowledge **representation language** is defined by two aspects:

1. Syntax: The syntax of a language defines which configurations of the components of the language constitute valid sentences.
2. Semantics: The semantics defines which facts in the world the sentences refer to, and hence the statement about the world that each sentence makes.

This is a very general idea, and not restricted to natural language.

Suppose the language is arithmetic, then „x", „^{3}" and „y" are components (or symbols or words) of the language the syntax says that „x ^{3} y" is a valid sentence in the language, but „^{3} ^{3} x y" is not the semantics say that „x ^{3} y" is false if y is bigger than x, and true otherwise. A good knowledge representation system for any particular domain should possess the following properties:

1. Representational Adequacy – the ability to represent all the different kinds of knowledge that might be needed in that domain.
2. Inferential Adequacy – the ability to manipulate the representational structures to derive new structures (corresponding to new knowledge) from existing structures.
3. Inferential Efficiency – the ability to incorporate additional information into the knowledge structure which can be used to focus the attention of the inference mechanisms in the most promising directions.
4. Acquisitional Efficiency – the ability to acquire new information easily. Ideally the agent should be able to control its own knowledge acquisition,

but direct insertion of information by a "knowledge engineer" would be acceptable.

In practice, the theoretical requirements for good knowledge representations can usually be achieved by dealing appropriately with a number of practical requirements:

1. The representations need to be complete – so that everything that could possibly need to be represented, can easily be represented.
2. They must be computable – implementable with standard computing procedures.
3. They should make the important objects and relations explicit and accessible – so that it is easy to see what is going on, and how the various components interact.
4. They should suppress irrelevant detail – so that rarely used details don't introduce necessary complications, but are still available when needed.
5. They should expose any natural constraints – so that it is easy to express how one object or relation influences another.
6. They should be transparent – so you can easily understand what is being said.
7. The implementation needs to be concise and fast – so that information can be stored, retrieved and manipulated rapidly.

3.2. Production Based System

A production system consists of four basic components:

1. A set of rules of the form Ci ® Ai where Ci is the condition part and Ai is the action part. The condition determines when a given rule is applied, and the action determines what happens when it is applied.
2. One or more knowledge databases that contain whatever information is relevant for the given problem. Some parts of the database may be permanent, while others may temporary and only exist during the solution of the current problem. The information in the databases may be structured in any appropriate manner.
3. A control strategy that determines the order in which the rules are applied to the database, and provides a way of resolving any conflicts that can arise when several rules match at once.
4. A rule applier which is the computational system that implements the control strategy and applies the rules.

Four Classes of Production Systems

1. A monotonic production system
2. A non-monotonic production system

3. A partially commutative production system
4. A commutative production system.

Advantages of Production Systems

1. Production systems provide an excellent tool for structuring AI programs.
2. Production Systems are highly modular because the individual rules can be added, removed or modified independently.
3. The production rules are expressed in a natural form, so the statements contained in the knowledge base should the a recording of an expert thinking out loud.

Disadvantages of Production Systems

One important disadvantage is the fact that it may be very difficult to analyse the flow of control within a production system because the individual rules don't call each other.

Production systems describe the operations that can be performed in a search for a solution to the problem. They can be classified as follows:

Monotonic Production System

A system in which the application of a rule never prevents the later application of another rule, that could have also been applied at the time the first rule was selected.

Partially Commutative Production System

A production system in which the application of a particular sequence of rules transforms state X into state Y, then any permutation of those rules that is allowable also transforms state X into state Y.

3.3. Inference

Two control strategies: forward chaining and backward chaining

Forward Chaining

Working from the facts to a conclusion. Sometimes called the datadriven approach. To chain forward, match data in working memory against 'conditions' of rules in the rule-base. When one of them fires, this is liable to produce more data. So the cycle continues.

Backward Chaining

Working from the conclusion to the facts. Sometimes called the goal-driven approach.

To chain backward, match a goal in working memory against 'conclusions' of rules in the rulebase.

3.3.1. Forward & Backward Chaining

* Here are two rules:

- ★ If corn is grown on poor soil, then it will get blackfly.
- ★ If soil hasn't enough nitrogen, then it is poor soil.
- ★ Forward chaining: This soil is low in nitrogen; therefore this is poor soil; therefore corn grown on it will get blackfly.
- ★ Backward chaining: This corn has blackfly; therefore it must have been grown on poor soil; therefore the soil must be low in nitrogen.

Properties of Forward Chaining

- ★ Sound and complete for first-order definite clauses: Datalog = first-order definite clauses + no functions
- ★ FC terminates for Datalog in finite number of iterations: May not terminate in general if α is not entailed. This is unavoidable: entailment with definite clauses is semidecidable.

Proof

Step 1

Step 2

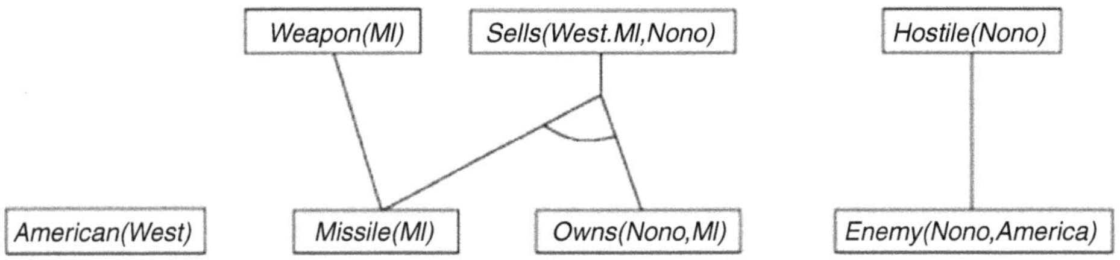

Step 3

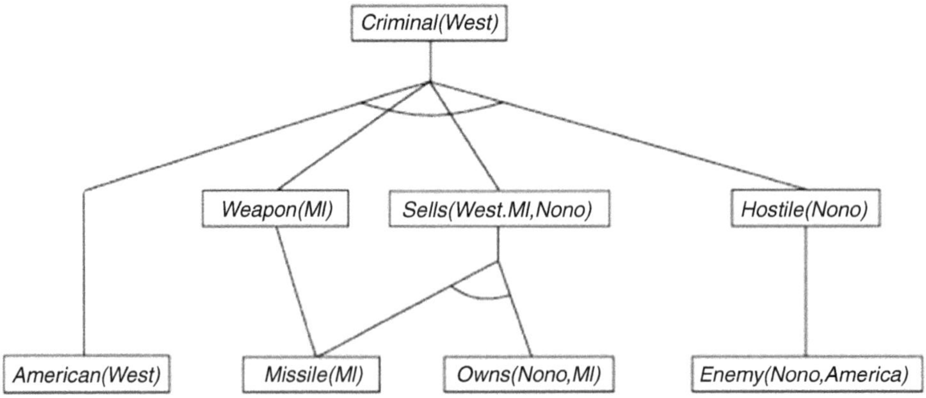

Backward Chaining

Step 1

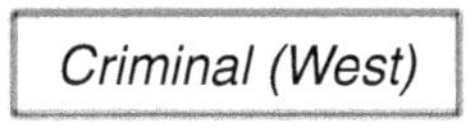

Step 2

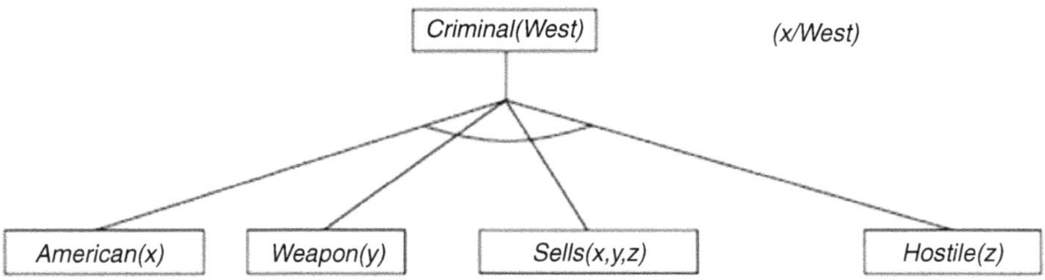

Step 3

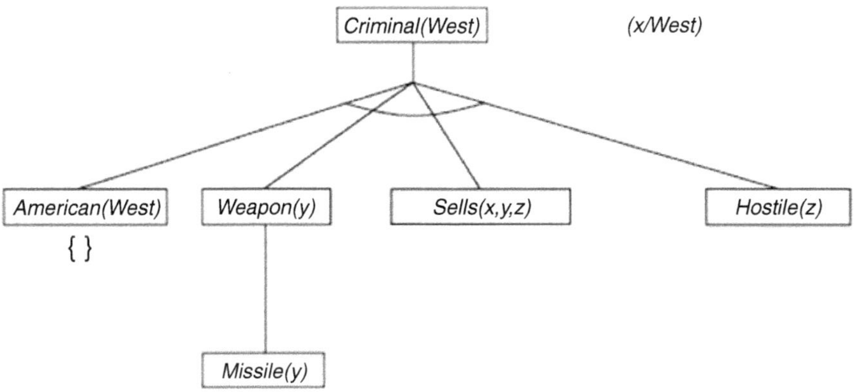

Step 4

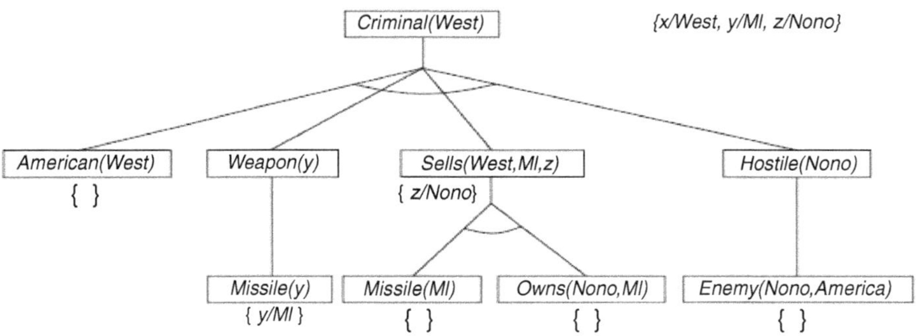

Properties of Backward Chaining

- ★ Depth-first recursive proof search: space is linear in size of proof. Incomplete due to infinite loops ⇒ fix by checking current goal against every goal on stack. Inefficient due to repeated subgoals (both success and failure).

  ⇒ fix using caching of previous results (extra space)
- ★ Widely used for logic programming.

3.4. Actions, Events and Situations

- ★ Reasoning about the outcome of actions is central to KB-agent.
- ★ How can we keep track of location in FOL?
  - Remember the multiple copies in PL.
- ★ Representing time by situations (states resulting from the execution of actions).
- ★ Situation calculus.

Situation Calculus

- ★ Actions are logical terms.
- ★ Situations are logical terms consiting of:
  - The initial situation I
  - All situations resulting from the action on I (=Result(a,s))
- ★ Fluents are functions and predicates that vary from one situation to the next.

  e.g., ¬Holding(G1, S0)
- ★ Eternal predicates are also allowed

  e.g., Gold(G1)

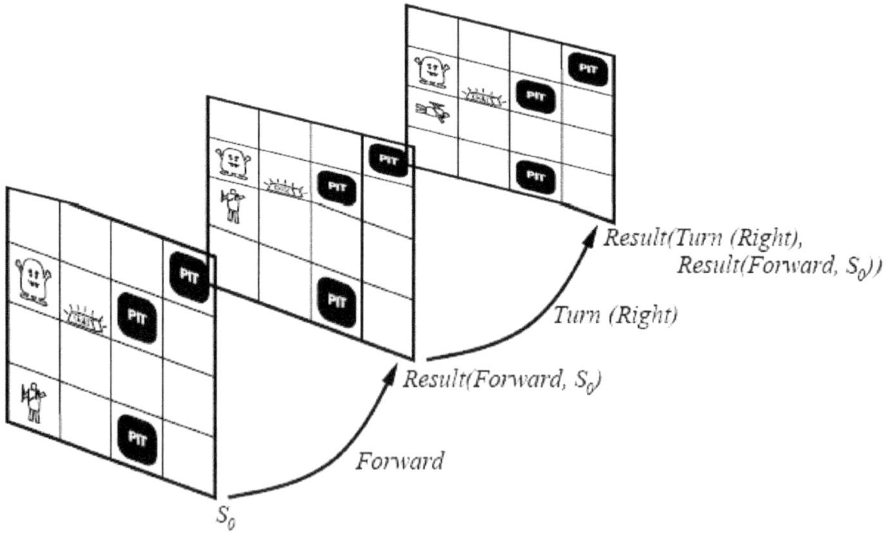

3.5. Mental Events and Objects

- ★ So far, KB agents can have beliefs and deduce new beliefs.
- ★ What about knowledge about beliefs?
- ★ What about knowledge about the inference process?
  - Requires a model of the mental objects in someone's head and the processes that manipulate those objects.
- ★ Relationships between agents and mental objects:

  believes, knows, wants,...
  - Believes(Lois,Flies(Superman)) with Flies(Superman) being a function ... a candidate for a mental object (reification).
- ★ Agent can now reason about the beliefs of agents.

3.6. Reasoning Systems for Categories

How to organize and reason with categories?

- ★ Semantic networks:
  - Visualize knowledge-base
  - Efficient algorithms for category membership inference
- ★ Description logics
  - Formal language for constructing and combining category definitions
  - Efficient algorithms to decide subset and superset relationships between categories.

3.7. Reasoning with Default Information

- ★ "The following courses are offered: ICS101, ICS110, ICS111, EE160".
- ★ Four (db):
  - Assume that this information is complete (not asserted ground atomic sentences are false).
  - = CLOSED WORLD ASSUMPTION
  - Assume that distinct names refer to distinct objects
  - = UNIQUE NAMES ASSUMPTION
- ★ Between one and infinity (logic):
  - Does not make these assumptions
  - Requires completion.

3.8. Fuzzy Logic

Fuzzy Logic (FL) is a method of reasoning that resembles human reasoning. The approach of FL imitates the way of decision making in humans that involves all intermediate possibilities between digital values YES and NO.

The conventional logic block that a computer can understand takes precise input and produces a definite output as TRUE or FALSE, which is equivalent to human's YES or NO.

The inventor of fuzzy logic, Lotfi Zadeh, observed that unlike computers, the human decision making includes a range of possibilities between YES and NO, such as –

- CERTAINLY YES
- POSSIBLY YES
- CANNOT SAY
- POSSIBLY NO
- CERTAINLY NO

The fuzzy logic works on the levels of possibilities of input to achieve the definite output.

Implementation

- ★ It can be implemented in systems with various sizes and capabilities ranging from small micro-controllers to large, networked, workstation-based control systems.
- ★ It can be implemented in hardware, software, or a combination of both.

Why Fuzzy Logic?

Fuzzy logic is useful for commercial and practical purposes:

- ★ It can control machines and consumer products.
- ★ It may not give accurate reasoning, but acceptable reasoning.
- ★ Fuzzy logic helps to deal with the uncertainty in engineering.

Fuzzy Logic Systems Architecture

It has four main parts as shown –

- ★ Fuzzification Module – It transforms the system inputs, which are crisp numbers, into fuzzy sets. It splits the input signal into five steps such as –
  - LP x is Large Positive
  - MP x is Medium Positive
  - S x is Small
  - MN x is Medium Negative
  - LN x is Large Negative
- ★ Knowledge Base – It stores IF-THEN rules provided by experts.
- ★ Inference Engine – It simulates the human reasoning process by making fuzzy inference on the inputs and IF-THEN rules.

- ✯ Defuzzification Module – It transforms the fuzzy set obtained by the inference engine into a crisp value.

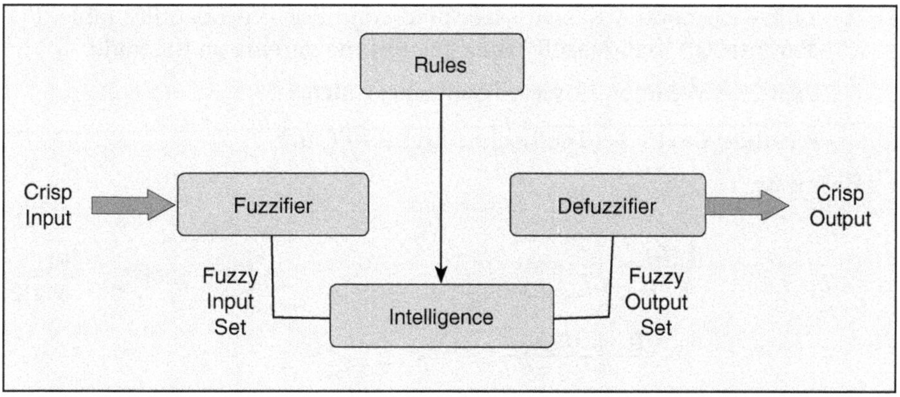

The membership functions work on fuzzy sets of variables.

Membership Function

Membership functions allow you to quantify linguistic term and represent a fuzzy set graphically. A membership function for a fuzzy set A on the universe of discourse X is defined as $\mu A:X \rightarrow [0,1]$.

Here, each element of X is mapped to a value between 0 and 1. It is called membership value or degree of membership. It quantifies the degree of membership of the element in X to the fuzzy set A.

- ✯ X axis represents the universe of discourse.
- ✯ Y axis represents the degrees of membership in the [0, 1] interval.

There can be multiple membership functions applicable to fuzzify a numerical value. Simple membership functions are used as use of complex functions does not add more precision in the output.

3.9. Bayesian Networks

- ✯ Represent dependencies among random variables.
- ✯ Give a short specification of conditional probability distribution.
- ✯ Many random variables are conditionally independent.
- ✯ Simplifies computations.
- ✯ Graphical representation.
- ✯ DAG – causal relationships among random variables.
- ✯ Allows inferences based on the network structure.

Definition of Bayesian Networks

A BN is a DAG in which each node is annotated with quantitative probability information, namely:

- ★ Nodes represent random variables (discrete or continuous).
- ★ Directed links X→Y: X has a direct influence on Y, X is said to be a parent of Y each node X has an associated conditional probability table, **P(Xi | Parents(Xi))** that quantify the effects of the parents on the node.

  Example: Weather, Cavity, Toothache, Catch

  Weather, Cavity → Toothache, Cavity → Catch

Example:

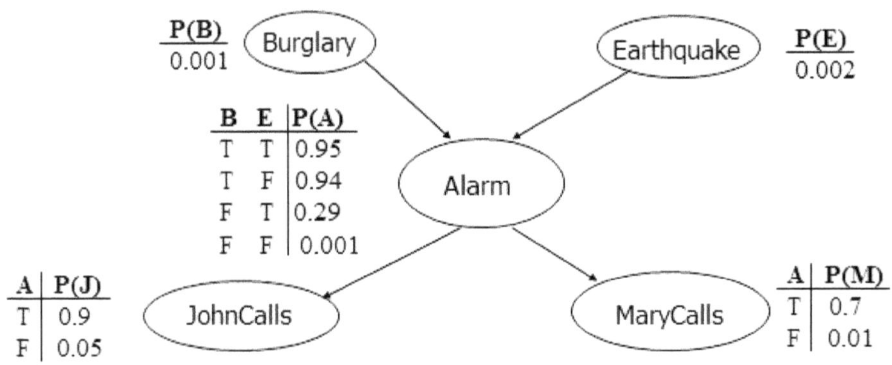

Conditional probability table

B	E	P(A | B, E) T	F
T	T	0.95	0.05
T	F	0.94	0.06
F	T	0.29	0.71
F	F	0.001	0.999

Bayesian Network Semantics

(A) Represent a probability distribution.

(B) Specify conditional independence – build the network.

Each value of the probability distribution can be computed as:

$P(X_1=x_1 \wedge \ldots X_n=x_n) = P(x1,\ldots, xn) = \prod i=1,n\ P(xi \mid Parents(xi))$

where Parents(xi) represent the specific values of Parents(Xi)

3.9.1. Building the Network

$P(X_1=x1 \wedge \ldots Xn=xn) = P(x1,\ldots, xn) =$

$P(xn \mid xn-1,\ldots, x1) * P(xn-1,\ldots, x1) = \ldots =$

$P(xn \mid xn-1,\ldots, x1) * P(xn-1 \mid xn-2,\ldots, x1) * \ldots P(x2 \mid x1) * P(x1) = \prod i=1,n\ P(xi \mid xi-1,\ldots, x1)$

We can see that $P(Xi \mid Xi-1,\ldots, X1) = P(xi \mid Parents(Xi))$ if $Parents(Xi) \subseteq \{ Xi-1,\ldots, X1\}$

- ✭ The condition may be satisfied by labeling the nodes in an order consistent with a DAG
- ✭ Intuitively, the parents of a node X_i must be all the nodes $X_{i-1}, \ldots, X_1$ which have a direct influence on X_i.
- ✭ Pick a set of random variables that describe the problem.
- ✭ Pick an ordering of those variables.
- ✭ While there are still variables repeat:
  - choose a variable X_i and add a node associated to X_i.
- ✭ Assign Parents(X_i) ← a minimal set of nodes that already exists in the network such that the conditional independence property is satisfied.
  - define the conditional probability table for X_i.
- ✭ Because each node is linked only to previous nodes → DAG
- ✭ P(MaryCalls | JohnCals, Alarm, Burglary, Earthquake) = P(MaryCalls | Alarm)

3.10. Dempster-Shafer Theory

- ✭ Dempster-Shafer theory is an approach to combining evidence.
- ✭ Dempster (1967) developed means for combining degrees of belief derived from independent items of evidence.
- ✭ His student, Glenn Shafer (1976), developed method for obtaining degrees of belief for one question from subjective probabilities for a related question.

People working in Expert Systems in the 1980s saw their approach as ideally suitable for such systems.

- ✭ Each fact has a degree of support, between 0 and 1:
  - 0 No support for the fact.
  - 1 full support for the fact.
- ✭ Differs from Bayesian approach in that:
  - Belief in a fact and its negation need not sum to 1.
  - Both values can be 0 (meaning no evidence for or against the fact).

3.10.1. Set of Possible Conclusions

$\Theta = \{ \theta 1 , \theta 2 , \ldots, \theta n \}$

Where:

- ✓ Θ is the set of possible conclusions to be drawn
- ✓ Each θi is mutually exclusive: at most one has to be true.
- ✓ Θ is Exhaustive: At least one θi has to be true.

3.10.2. Frame of Discernment

$\Theta = \{ \theta 1 , \theta 2 , \ldots, \theta n \}$

Bayes was concerned with evidence that supported single conclusions (*e.g.*, evidence for each outcome θi in Θ): p(θi| E).

D-S Theory is concerned with evidences which support subsets of outcomes in Θ, *e.g.*, θ1 v θ2 v θ3 or {θ1, θ2, θ3}.

The "frame of discernment" (or "Power set") of Θ is the set of all possible subsets of Θ:

e.g., if Θ = {θ1, θ2, θ3}.

Then the frame of discernment of Θ is:

(Ø, θ1, θ2, θ3, {θ1, θ2}, {θ1, θ3}, {θ2, θ3}, { θ1, θ2, θ3})

- Ø, the empty set, has a probability of 0, since one of the outcomes has to be true.
- Each of the other elements in the power set has a probability between 0 and 1.
- The probability of {θ1, θ2, θ3} is 1.0 since one has to be true.

3.10.3. Belief Intervals & Probability

The probability in A falls somewhere between bel(A) and pl(A).

- bel(A) represents the evidence we have for A directly So prob(A) cannot be less than this value.
- pl(A) represents the maximum share of the evidence we could possibly have, if, for all sets that intersect with A, the part that intersects is actually valid. So pl(A) is the maximum possible value of prob(A).

A	{B}	{J}	{S}	{B,J}	{B,S}	{J,S}	{B,J,S}
m(A)	0.1	0.2	0.1	0.1	0.1	0.3	0.1
bel(A)	0.1	0.2	0.1	0.4	0.3	0.6	1.0
pl(A)	0.4	0.7	0.6	0.9	0.8	0.9	1.0

3.10.4. Belief Intervals

Belief intervals allow Demspter-Shafer theory to reason about the degree of certainty or certainty of our beliefs.

- A small difference between belief and plausibility shows that we are certain about our belief.
- A large difference shows that we are uncertain about our belief. However, even with a 0 interval, this does not mean we know which conclusion is right. Just how probable it is!

A	{B}	{J}	{S}	{B,J}	{B,S}	{J,S}	{B,J,S}
m(A)	0.1	0.2	0.1	0.1	0.1	0.3	0.1
bel(A)	0.1	0.2	0.1	0.4	0.3	0.6	1.0
pl(A)	0.4	0.7	0.6	0.9	0.8	0.9	1.0

3.11. Rule Based Approach

- ☆ One can often represent the expertise that someone uses to do an expert task as rules.
- ☆ A rule means a structure which has an if component and a then component.
- ☆ This is actually a very old idea indeed.

Rules

Examples

if - the leaves are dry, brittle and discoloured

then - the plant has been attacked by red spider mite

if - the customer closes the account

then - delete the customer from the database

- The statement, or set of statements, after the word if represents some pattern which you may observe.
- The statement, or set of statements, after the word then represents some conclusion that you can draw, or some action that you should take.

A Rule-based System, Therefore, Either

- identifies a pattern and draws conclusions about what it means,
- identifies a pattern and advises what should be done about it,
- identifies a pattern and takes appropriate action.

4

Planning and Machine Learning

4.1. Planning with State Space Search

The agent first generates a goal to achieve and then constructs a plan to achieve it from the current state.

Problem Solving to Planning

Representation Using Problem Solving Approach

- ✓ Forward search
- ✓ Backward search
- ✓ Heuristic search

Representation Using Planning Approach

- ✓ STRIPS-standard research institute problem solver.
- ✓ Representation for states and goals.
- ✓ Representation for plans.
- ✓ Situation space and plan space.
- ✓ Solutions.

Why Planning?

Intelligent agents must operate in the world. They are not simply passive reasons (Knowledge Representation, reasoning under uncertainty) or problem solvers (Search), they must also act on the world.

We want intelligent agents to act in "intelligent ways". Taking purposeful actions, predicting the expected effect of such actions, composing actions together to achieve complex goals. For example if we have a robot we want robot to decide what to do; how to act to achieve our goals.

Planning Problem

How to change the world to suit our needs

- ✯ Critical issue: we need to reason about what the world will be like after doing a few actions, not just what it is like now GOAL: Craig has coffee.
- ✯ CURRENTLY: robot in mailroom, has no coffee, coffee not made, Craig in office etc.
- ✯ TO DO: goto lounge, make coffee

Partial Order Plan

- ✯ A partially ordered collection of steps:
  - *Start step* has the initial state description and its effect.
  - *Finish step* has the goal description as its precondition.
  - *Causal links* from outcome of one step to precondition of another step.
  - *Temporal ordering* between pairs of steps.
- ✯ An open condition is a precondition of a step not yet causally linked.
- ✯ A plan is ***complete*** if every precondition is achieved.
- ✯ A precondition is ***achieved*** if it is the effect of an earlier step and no possibly intervening step undoes it.

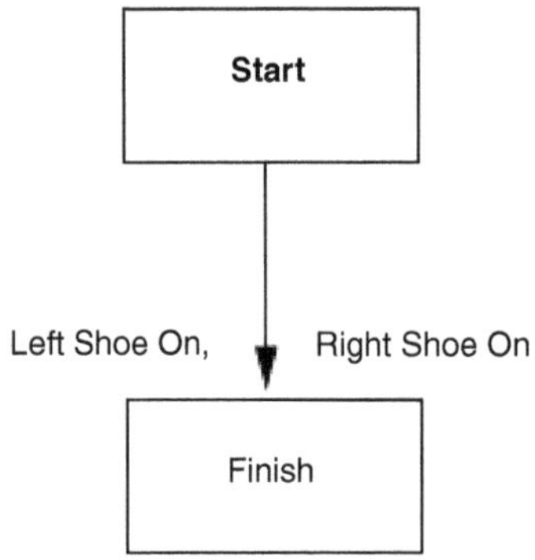

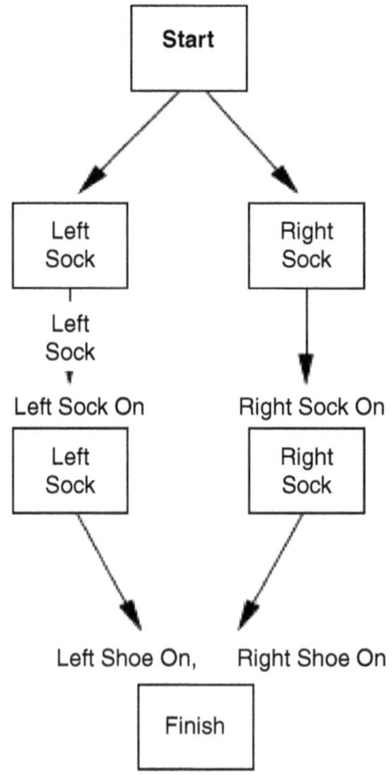

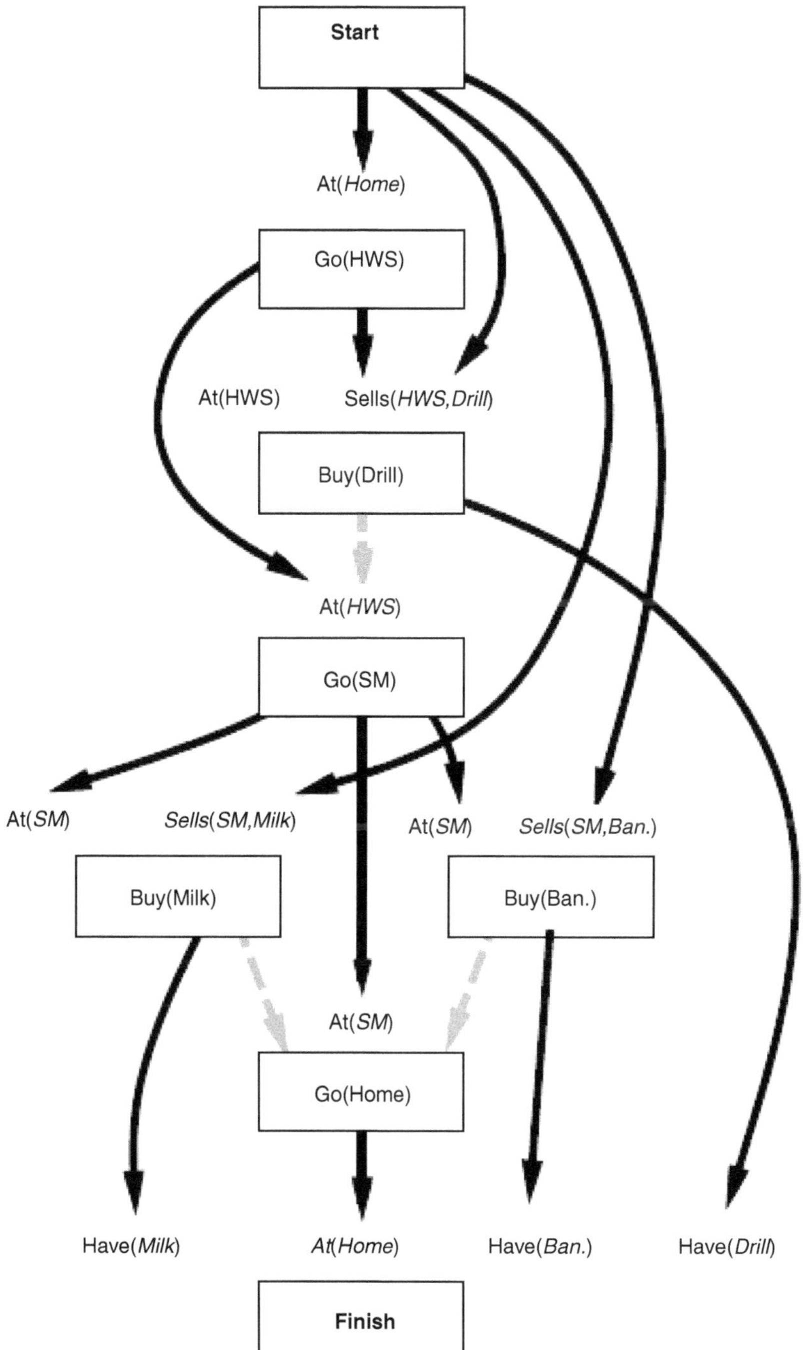
Start
At(Home)
Go(HWS)
At(HWS)
Sells(HWS,Drill)
Buy(Drill)
At(HWS)
Go(SM)
At(SM)
Sells(SM,Milk)
At(SM)
Sells(SM,Ban.)
Buy(Milk)
Buy(Ban.)
At(SM)
Go(Home)
Have(Milk)
At(Home)
Have(Ban.)
Have(Drill)
Finish

Partial Order Plan Algorithm

```
function POP(initial, goal, operators) returns plan
    plan ← MAKE-MINIMAL-PLAN(initial, goal)
    loop do
        if SOLUTION?(plan) then return plan
        S_need, c ← SELECT-SUBGOAL(plan)
        CHOOSE-OPERATOR(plan, operators, S_need, c)
        RESOLVE-THREATS(plan)
    end

function SELECT-SUBGOAL(plan) returns S_need, c
    pick a plan step S_need from STEPS(plan)
        with a precondition c that has not been achieved
    return S_need, c
```

```
procedure CHOOSE-OPERATOR(plan, operators, S_need, c)
    choose a step S_add from operators or STEPS(plan) that has c as an effect
    if there is no such step then fail
    add the causal link S_add --c--> S_need to LINKS(plan)
    add the ordering constraint S_add ≺ S_need to ORDERINGS(plan)
    if S_add is a newly added step from operators then
        add S_add to STEPS(plan)
        add Start ≺ S_add ≺ Finish to ORDERINGS(plan)

procedure RESOLVE-THREATS(plan)
    for each S_threat that threatens a link S_i --c--> S_j in LINKS(plan) do
        choose either
            Demotion: Add S_threat ≺ S_i to ORDERINGS(plan)
            Promotion: Add S_j ≺ S_threat to ORDERINGS(plan)
        if not CONSISTENT(plan) then fail
    end
```

4.2. Stanford Research Institute Problem Solver (STRIPS)

STRIPS is a classical planning language, representing plan components as states, goals, and actions, allowing algorithms to parse the logical structure of the planning problem to provide a solution.

In STRIPS, state is represented as a conjunction of positive literals. Positive literals may be a propositional literal (*e.g.*, Big ^ Tall) or a first-order literal (*e.g.*, At(Billy, Desk)). The positive literals must be grounded – may not contain a variable (*e.g.*, At(x, Desk)) – and must be function-free – may not invoke a function to calculate a value (*e.g.*, At(Father(Billy), Desk)).

Any state conditions that are not mentioned are assumed false.

The goal is also represented as a conjunction of positive, ground literals. A state satisfies a goal if the state contains all of the conjuncted literals in the goal; *e.g.*, Stacked ^ Ordered ^ Purchased satisfies Ordered ^ Stacked.

Actions (or operators) are defined by action schemas, each consisting of three parts:

* The action name and any parameters.
* Preconditions which must hold before the action can be executed. Preconditions are represented as a conjunction of function-free, positive literals. Any variables in a precondition must appear in the action's parameter list.
* Effects which describe how the state of the environment changes when the action is executed. Effects are represented as a conjunction of function-free literals. Any variables in a precondition must appear in the action's parameter list. Any world state not explicitly impacted by the action schema's effect is assumed to remain unchanged.

The following, simple action schema describes the action of moving a box from location x to location y:

Action: *MoveBox*(x, y)

Precond: *BoxAt*(x)

Effect: *BoxAt*(y), $\neg$ *BoxAt*(x)

If an action is applied, but the current state of the system does not meet the necessary preconditions, then the action has no effect. But if an action is successfully applied, then any positive literals, in the effect, are added to the current state of the world; correspondingly, any negative literals, in the effect, result in the removal of the corresponding positive literals from the state of the world.

4.2.1. STRIPS Planning Algorithm

As previously referenced, STRIPS began as an automated planning algorithm and has double meaning to describe the language (described above) used to provide input to that algorithm. While the algorithm does not scale well to real-world problems, it, like the language, serves as a foundational starting point to developing and understanding more powerful automated planning algorithms. The STRIPS algorithm [3] is found below, followed by a brief commentary:

STRIPS(A, s, g)

p = empty plan

loop...

if s satisfies g then return p

a = [an applicable action in A, relevant for g]

if a = null, then return failure

$p' = \text{STRIPS}(A, s, \text{precond}(a))$

if p' = failure, then return failure

s = apply p' to s

s = apply a to s

$p = p + p' + a$

In the above STRIPS algorithm, A represents all of the possible, grounded actions (*i.e.*, action schemas with variables replaced with values), while s is the current state, and g is the goal state.

Critically important to note is that this algorithm is a backward-search algorithm. In other words, the goal state of the planning problem is fed into the algorithm as s, while the initial state of the planning problem is provided as g. The returned solution, if any, simply needs to be reversed to provide the solution to the planning problem. Simply put, it works backwards from the goal to figure out a logical path to get there.

Explanation

Most expert systems have explanation facilities that allow the user to ask questions- why and how it reached some conclusion. The questions are answered by referring to the system goals, the rules being used, and existing problem solving. The rules typically reflect empirical or "compiled" knowledge. They are codes of an expert's rule of thumb, not the expert's deeper understanding.

Example

Dialog with an expert system designed to give advice on car problems.

System Does the car start?

User No.

System Does the engine turn over?

User yes

System Do you smell gas?

User yes

System Recommendation-wait 5 minutes and try again.

User Why?

System I used the rule:

If not start, and engine_turn_over, and smell_gas

Then recommendation is "wait 5 minutes and try again".

User

Note: The rule gives the correct advice for a flooded car. And knows the questions to be ask to determine if the car is flooded, but it does not contain the knowledge of what a flooded car is and why waiting will help.

Types of Explanation

There are four types of explanations commonly used in expert systems.

- ✓ Rule trace reports on the progress of a consultation;
- ✓ Explanation of how the system reached to the give conclusion;
- ✓ Explanation of why the system did not give any conclusion.
- ✓ Explanation of why the system is asking a question;

Learning

4.3. Machine Learning

- Like human learning from past experiences,a computer does not have "experiences".
- A computer system learns from data, which represent some "past experiences" of an application domain.
- Objective of machine learning : learn a target function that can be used to predict the values of a discrete class attribute, *e.g.*, approve or not-approved, and high-risk or low risk.
- The task is commonly called: **Supervised learning, classification, or inductive learning.**

4.3.1. Forms of Learning

Learning is essential for unknown environments, *i.e.*, when designer lacks omniscience

- Learning is useful as a system construction method:
    - *i.e.*, expose the agent to reality rather than trying to write it down.
    - Learning modifies the agent's decision mechanisms to improve performance.

Learning agents

Design of a learning element based on four major factors:

– Which component is to be improved

– What prior knowledge the agent already has

– What feedback is available to learn from

– What representation is used for the component

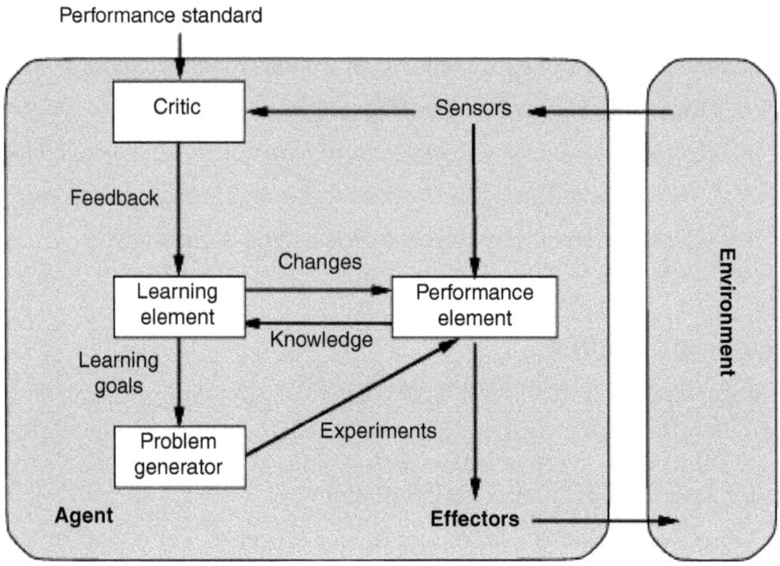

Components to be learned

- ★ Direct mapping from conditions
- ★ Infer relevant properties of the world
- ★ Information about the way the world evolves
- ★ Utility information
- ★ Action value information
- ★ Goals – classes of states
    - *e.g*: An agent training to become a taxi driver.
    - Learn a condition – action rule for when to brake
    - Learn the instructor instruction
    - Observe the images around the environment

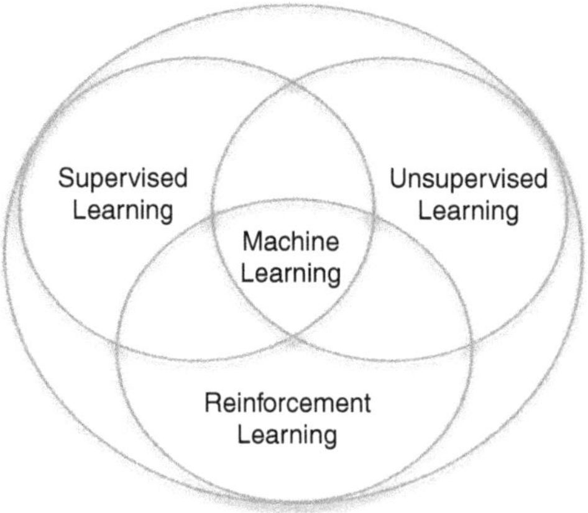

Types of Learning

- ★ Supervised learning: correct answer for each example. Answer can be a numeric variable, categorical variable, etc.
- ★ Unsupervised learning: correct answers not given – just examples (*e.g.*, the same figures as above , without the labels).
- ★ Reinforcement learning: rewards from sequence of actions.

4.3.2. Supervised Machine Learning

- ★ Supervised learning is where you have input variables (X) and an output variable (Y) and you use an algorithm to learn the mapping function from the input to the output.

  Y = f(X)

- ★ The goal is to approximate the mapping function so well that when you have new input data (X) that you can predict the output variables (Y) for that data.
- ★ Supervised learning problems can be further grouped into:
  - Classification: A classification problem is when the output variable is a category, such as "red" or "blue" or "disease" and "no disease".
  - Regression: A regression problem is when the output variable is a real value, such as "dollars" or "weight".

Unsupervised Machine Learning

- ★ Unsupervised learning is where you only have input data (X) and no corresponding output variables.
- ★ It can be further grouped into clustering and association problems.
  - Clustering: to discover the inherent groupings in the data, such as grouping customers by purchasing behaviour.
  - Association: to discover rules that describe large portions of your data, such as people that buy X also tend to buy Y.

Semi-Supervised Machine Learning

- ★ Problems where you have a large amount of input data (X) and only some of the data is labeled (Y) are called semi-supervised learning problems.
- ★ These problems sit in between both supervised and unsupervised learning.
  - A good example is a photo archive where only some of the images are labeled and the majority are unlabeled.

Reinforcement Learning

- ★ feedback concerning the desirability of the agent's behavior is available.
- ★ may not be directly attributable to a particular action.
- ★ feedback may occur only after a sequence of actions.
- ★ the agent or component knows that it did something right (or wrong), but not what action caused it.

4.3.3. Decision Tree

- ★ Decision Tree Learning (DTL) is a form of inductive learning task, meaning it has the following objective:
  - Use a training set of examples to create a hypothesis that makes general conclusions.

- Attribute: a variable that we take into account in making a decision.
- Target attribute: the attribute that we want to take on a certain value, we'll decide based on it.
- Decision Trees are an elegant way of learning – it is easy to expose their logic and understand what it has learned.
- Decision Trees are not always the best way to learn – they have some weaknesses.
- But it also has its own set of strengths.

Example

Problem

Decide whether to wait for a table at a restaurant, based on the following attributes:

1. Alternate: is there an alternative restaurant nearby?
2. Fri/Sat: is today Friday or Saturday?
3. Hungry: are we hungry?
4. Patrons: number of people in the restaurant (None, Some, Full)
5. Price: price range ($, $$, $$$)
6. Raining: is it raining outside?
7. Reservation: have we made a reservation?
8. Type: kind of restaurant (French, Italian, Chinese)
9. Wait Estimate: estimated waiting time (0-10, 10-30, 30-60, >60)

One possible representation for hypotheses

e.g., here is the "true" tree for deciding whether to wait:

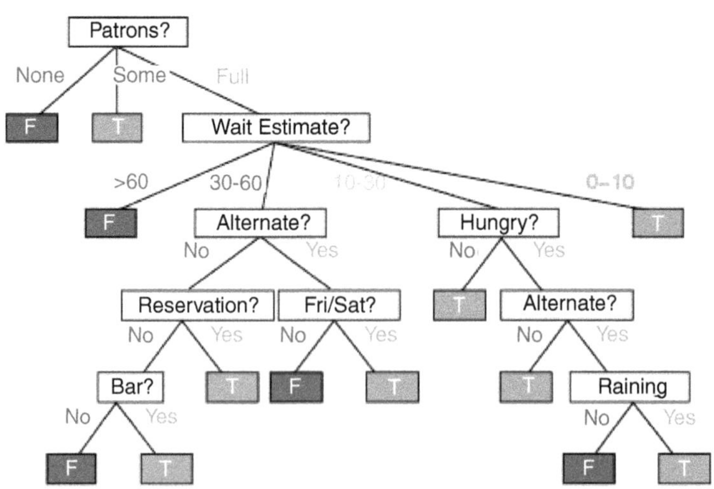

Expressiveness of Decision Trees

- ★ Decision trees can also be expressed in logic as implication sentences.
- ★ They can express propositional logic sentences.
  - each row in the truth table of a sentence can be represented as a path in the tree.
  - often there are more efficient trees.
- ★ some functions require exponentially large decision trees parity function, majority function

4.3.4. Decision Tree Learning

- ★ Aim: find a small tree consistent with the training examples.
- ★ Idea: (recursively) choose "most significant" attribute as root of (sub)tree

```
function DTL(examples, attributes, default) returns a decision tree
  if examples is empty then return default
  else if all examples have the same classification then return the classification
  else if attributes is empty then return MODE(examples)
  else
      best ← CHOOSE-ATTRIBUTE(attributes, examples)
      tree ← a new decision tree with root test best
      for each value v_i of best do
          examples_i ← {elements of examples with best = v_i}
          subtree ← DTL(examples_i, attributes − best, MODE(examples))
          add a branch to tree with label v_i and subtree subtree
      return tree
```

4.4. Probability Basics

- ★ A well-known and well-understood framework for uncertainty.
- ★ Clear semantics.
- ★ Provides principled answers for:
  - Combining evidence
  - Predictive & Diagnostic reasoning
  - Incorporation of new evidence
- ★ Intuitive (at some level) to human experts
- ★ Can be learned

4.4.1. Bayesian Viewpoint

- ★ Probability is "degree-of-belief", or "degree-of-uncertainty".
- ★ To the Bayesian, probability lies subjectively in the mind, and can–with validity–be different for people with different information
- ★ For example, the probability that Wayne will get rich from selling his kidney.

★ In contrast, to the frequentist, probability lies objectively in the external world.

★ The Bayesian viewpoint has been gaining popularity in the past decade, largely due to the increase computational power that makes many of the calculations that were previously intractable, feasible.

Bayes' Rule

$P(A \wedge B) = P(A|B)\,P(B) = P(B|A)\,P(A)$

Bayes Theorem Example:

★ Bayes Theorem is a very common and fundamental theorem used in Data mining and Machine learning.

★ Its formula is pretty simple:

★ P(X|Y) = (P(Y|X) * P(X)) / P(Y)

★ which is Posterior = (Likelihood * Prior) / Evidence

4.4.2. Bayesian Networks

★ More commonly called *graphical models.*

★ A way to depict conditional independence relationships between random variables.

A compact specification of full joint distributions

★ **Nodes:** random variables

- Can be assigned (observed) or unassigned (unobserved).

★ **Arcs:** interactions

- An arrow from one variable to another indicates direct influence.
- Encode conditional independence:
  - *Weather* is independent of the other variables
  - *Toothache* and *Catch* are conditionally independent given *Cavity*
  - Must form a directed, *acyclic* graph

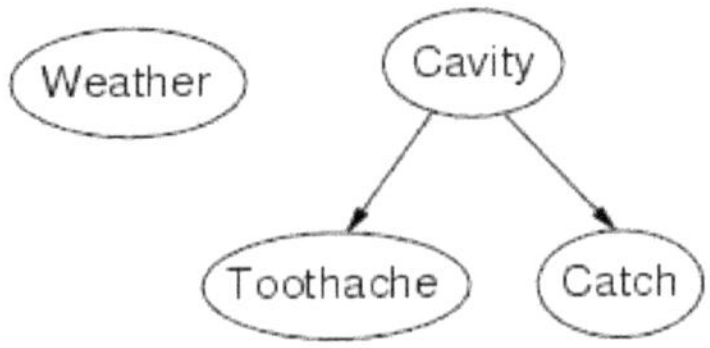

Expressiveness of Formal Models

★ Atomic models: each possible world is an atom or token, i.e., no internal structure, *e.g.*, Heads or Tails.

★ Propositional models: each possible world defined by values assigned to variables. *e.g.*, Boolean logic, Bayes net, neural net.

★ First-order models: each possible world defined by objects and relations.

Dynamic Bayesian Networks

- ☆ It is a kind of BN that represents a temporal probability model.
- ☆ It is very clear that every HMM can be represented as a DBN with a single state variable and a single evidence variable.
- ☆ It is also the case that every discrete variable DBN can be represented as an HMM.

4.4.3. Semantics and Inference of Bayesian Networks

- ☆ Semantics
  - Represent Full Joint Distributions
  - Represent Conditional Distributions
- ☆ Inference
  - Exact Inference
  - Approximate Inference

Sampling Methods

- ☆ Relational and First Order Probability Models.
- ☆ Observation: the world consists of many distinct entities with similar behaviours.
- ☆ Exploit this redundancy to make our models simpler.
- ☆ This was the idea of FOL: use quantification to eliminate redundant sentences over ground literals.

Probabilistic Reasoning over Time

- ☆ Belief State: Represents which state of the world currently possible.
- ☆ Agent can predict how the world might evolve in the next time step using belief state and transition model.
- ☆ Agent can update the belief state from the percepts observed and a sensor model.
- ☆ Transition and sensor model may be uncertain.
- ☆ Transition model describes the probability distribution of the variable at time *t*, given the state of the world at past times.
- ☆ Sensor Model describes the probability of each percept at time *t*, given the current state.

Time and Uncertainty

- ☆ Dynamic aspects of the problems are essential

Example

- ☆ Blood sugar level and measurement can change over time depending on:
  - Recent food intake.

- Insulin doses.
- Metabolic activity.

★ To assess the current state from the history of evidence.

★ To predict the outcomes of treatment actions.

★ We must model these changes.

★ Other Examples

- Tracking the location of a robot.
- Tracking the economic activity of a nation.

States and Observations

Model Dynamic Changes

★ X_t- set of state variables at time t

- It is assumed to be unobservable

★ E_t- set of observable evidence variable

★ $E_t=e_t$ for some set of values e_t

★ The interval between the time slices also depends on the problem.

★ Example: for diabetes monitoring

- A suitable interval might be an hour rather than a day.

Transition and Sensor Models

★ Need to specify how the world evolves and how the evidence variables get their values.

★ Transition Model is represented as:

- $P(X_t | X_{0:t-1})$

★ When t increases $X_{0:t-1}$ is unbounded.

★ So we use Markov assumption.

Markov Assumption

★ Current state depends on only a finite number of previous states.

★ Processes that satisfy this assumption are called Markov processes / Markov Chains.

We use first Order Markov Process

★ Current state depends only on the previous state and not on any earlier states.

Transition Model

★ First order Markov Process:

- $P(X_t | X_{0:t-1})=P(X_t | X_{t-1})$

★ Second order Markov Process:

- $P(X_t | X_{0:t-1}) = P(X_t | X_{t-2}, X_{t-1})$

Sensor Model (Observation Model)

★ The evidence variables E_t could depend on previous variables as well as the current state variables.

★ $P(E_t | X_{0:t}, E_{0:t-1}) = P(E_t | X_t)$

Inference in Temporal Models

★ Filtering

★ Prediction

★ Smoothing

★ Most likely explanation

Filtering

★ Task of computing the Belief State.

- *i.e.*, Posterior distribution over the most recent date given all evidence to date.
- Aka. State Estimation

★ Need to compute $P(X_t | e_{1:t})$.

Example

Computing the probability of rain today, given all the observations of the umbrella carrier made so far.

Filtering is what a rational agent does to keep track of the current state so that rational decisions can be made.

Prediction

Task of computing the posterior distribution over future state, given all evidence to date.

i.e., Need to compute $P(X_{t+k} | e_{1:t})$ for some $k>0$.

Example

Computing the probability of rain three days from now, given all the observations to date.

Here $k=3$.

Useful for evaluating possible courses of actions based on their expected outcomes.

Smoothing

Task of Computing the posterior distribution over a past state, given all evidence up to the present.

Need to compute $P(X_k | e_{1:t})$ for some k such that 0<=k<=t.

Example

Computing the probability that it rained last Wednesday, given all the observations of the umbrella carrier made up to today.

Useful for a better estimate of the state than was available at the time, because it incorporates more evidence.

Most likely Explanations

Given a sequence of observations, Find the sequence of states that is most likely to have generated those observations.

i.e., Compute $\text{argmax}_{x1:t} P(x_{1:t} | e_{1:t})$

Example

If the umbrella appears on each of the three days and is absent on the fourth, then the most likely explanation is that it rained on the first three days and did not rain on the fourth.

Application

Speech Recognition

Find the most likely sequence of words, given a series of sounds and the reconstructions of bit strings transmitted over a noisy channel.

4.5. Hidden Markov Model

- ★ Temporal probabilistic model.
- ★ State of the process described by a single discrete random variable.
- ★ Possible values of the variable are the possible state of the world.

Example

In Umbrella World, $Rain_t$ is the random variable.

Using X_t (state variable):

- Can give concrete form to the representation of
- Transition model
- Sensor model and
- Forward and Backward messages.

Transition Model

- ★ Let the state variable X_t have values denoted by integers 1,....S, where S is the number of possible states.

- The transition model $P(X_t | X_{t-1})$ becomes an SXS matrix T, where
- $T_{ij}=P(X_t=j | X_{t-1}=i)$
- T_{ij} is the probability of transition from state i to state j.

Sensor Model

- Values of the evidence variable E_t is known at time t (e_t)
- Need to specify how likely it is that the state causes e_t to appear.
- *i.e.*, $P(e_t | X_t=i)$ for each state i.

Dynamic Bayesian Networks vs Hidden Markov Model

Decomposing the state of a complex system into its constituent variables, DBN can take advantage of sparseness in the temporal probability model.

For Example

If DBN has 20 Boolean state variables, each of which has 3 parents in the preceding slice, then,

- DBN transition model has $20X2^3 =160$ probabilities
- But HMM has 20^{20}.

Drawbacks in Hidden Markov Model

- Requires much more space.
- Huge transition matrix makes HMM inference much more expensive.
- Problem of learning such a huge number of parameters makes the pure HMM model unsuitable for large problems.

4.6. Artificial Neural Networks

- Neurons and the Brain
- Neural Networks
- Perceptrons
- Multi-layer Networks
- Applications

Neural Networks

- complex networks of simple computing elements.
- capable of learning from examples:
  - with appropriate learning methods
- collection of simple elements performs high-level operations:
  - thought
  - reasoning
  - consciousness.

Artificial Neuron Diagram

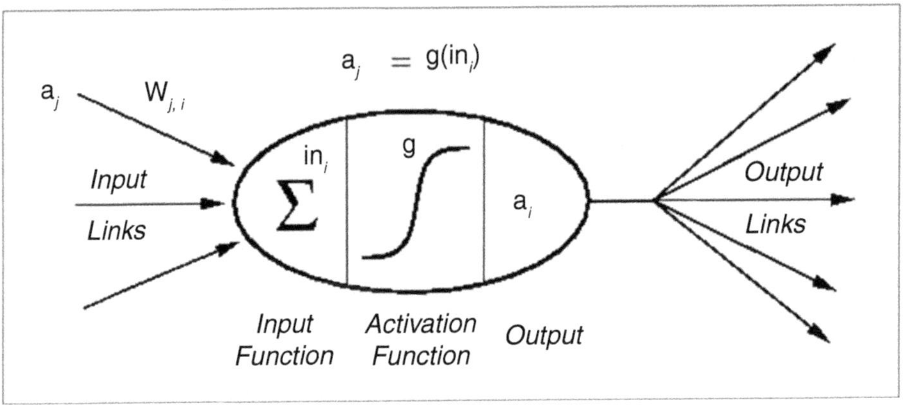

- ★ weighted inputs are summed up by the input function.
- ★ the (nonlinear) activation function calculates the activation value, which determines the output.

Common Activation Functions

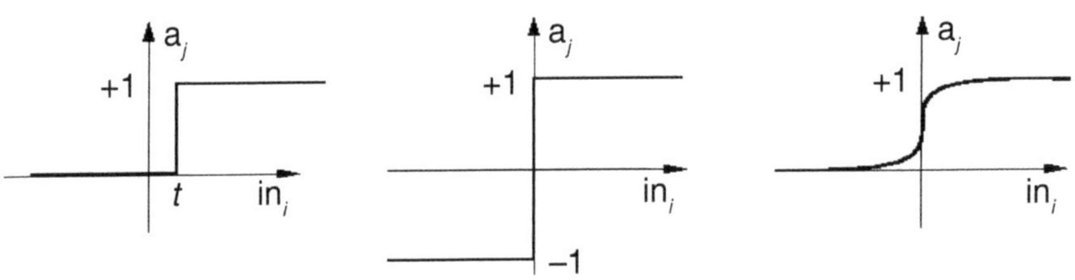

(a) Step function (b) Sign function (c) Sigmoid function

$Step_t(x)$	=	1	if x >= t, else 0
Sign(x)	=	+1	if x >= 0, else –1
Sigmoid(x)	=	$1/(1+e^{-x})$	

Neural Networks and Logic Gates

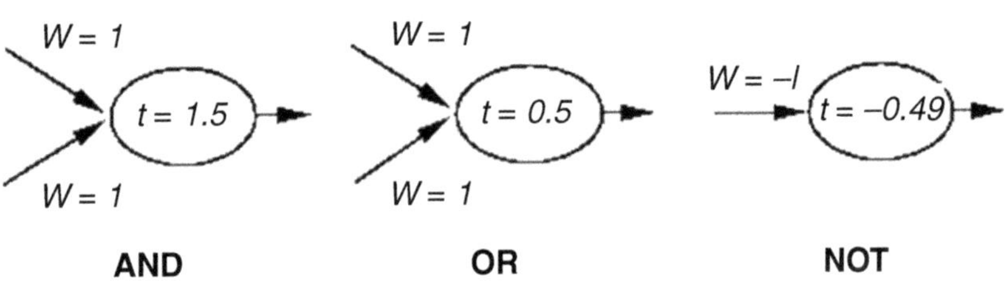

- ★ Simple neurons with can act as logic gates:
  - appropriate choice of activation function, threshold, and weights.
  - step function as activation function.

Network Structures

- In principle, networks can be arbitrarily connected:
  - occasionally done to represent specific structures:
    - – semantic networks.
    - – logical sentences.
  - makes learning rather difficult.
- Layered structures:
  - networks are arranged into layers.
  - interconnections mostly between two layers.
  - some networks have feedback connections.

Perceptrons

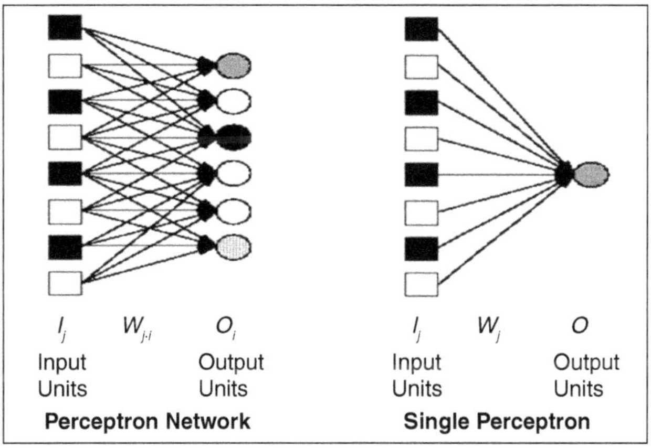

- Single layer, feed-forward network.
- Historically one of the first types of neural networks.
- the output is calculated as a step function applied to the weighted sum of inputs.
  - linearly separable.

4.7. Non-Parametric Models

- Doesn't mean that the model lacks parameters.
- Parameters are not known or fixed in advance.
- Make no assumptions about probability distributions.
- Structure determined from the data:
  - Nearest neighbour models.
  - Finding nearest neighbours with kd trees.
  - Locality-sensitive hashing.
  - Non-parametric regression .

Comparison of Models

Parametric Models	Non-Parametric Models
• data summarized by a fixed set of parameters • once learned, the original data can be discarded • good when data set is relatively small – avoids overfitting • best when correct parameters are chosen!	• data summarized by an unknown (or non-fixed) set of parameters • must keep original data to make predictions or to update model • may be slower, but generally more accurate

Nearest Neighbour Models

- ✯ Given query xq answer query by finding the *k examples.*
  - nearest to xq.
- ✯ Classification: take plurality vote (majority for binary classification) of neighbours.
- ✯ Regression take mean or median of neighbour values.

4.8. Support Vector Machine (SVM)

- ✯ SVMs maximize the *margin* around the separating hyperplane:
  - A.k.a. large margin classifiers.
- ✯ The decision function is fully specified by a subset of training samples, *the support vectors.*
- ✯ Solving SVMs is a *quadratic programming* problem.
- ✯ Seen by many as the most successful current text classification method*

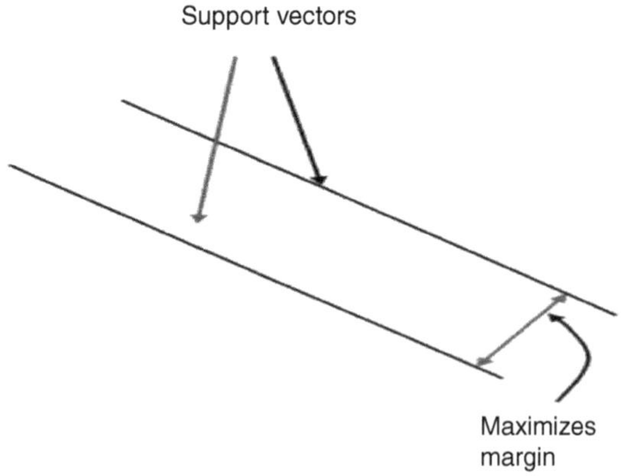

Properties of SVM

- ★ Flexibility in choosing a similarity function.
- ★ Sparseness of solution when dealing with large data sets.
- ★ Ability to handle large feature spaces.
- ★ Over fitting can be controlled by soft margin approach.
- ★ Nice math property: a simple convex optimization problem which is guaranteed to converge to a single global solution.
- ★ Feature Selection.

Explanation-based learning

- ★ A method to extract general rules from individual observations.
- ★ The goal is to *solve a similar problem faster next time.*
- ★ Memoization - speed up by saving results and avoiding solving a problem from scratch.
- ★ EBL does it one step further - from observations to rules.

Why EBL?

- ★ Explaining why something is a good idea is *much easier* than coming up with the idea.
- ★ Once something is understood, it can be generalized and reused in other circumstances.
- ★ Extracting general rules from examples.
- ★ EBL constructs two proof trees simultaneously by *variablization* of the constants in the first tree.

Learning Using Relevant Information

- ★ Prior knowledge: People in a country usually speak the same language:

  Nat(x,n) ^Nat(y,n)^Lang(x,l)=>Lang(y,l)
- ★ Observation: Given nationality, language is fully determined:
  - Given Fernando is Brazilian & speaks Portuguese

    Nat(Fernando,B) ^ Lang(Fernando,P)
- ★ We can logically conclude:
  - Nat(y,B) => Lang(y,P)

Reinforcement Learning

- ★ RL deals with agents that must sense & act upon their environment.
- ★ This is combines classical AI and machine learning techniques.
- ★ It the most comprehensive problem setting.

Examples

- A robot cleaning my room and recharging its battery.
- Robot-soccer.
- How to invest in shares.
- Modeling the economy through rational agents.
- Learning how to fly a helicopter.
- Scheduling planes to their destinations and so on.

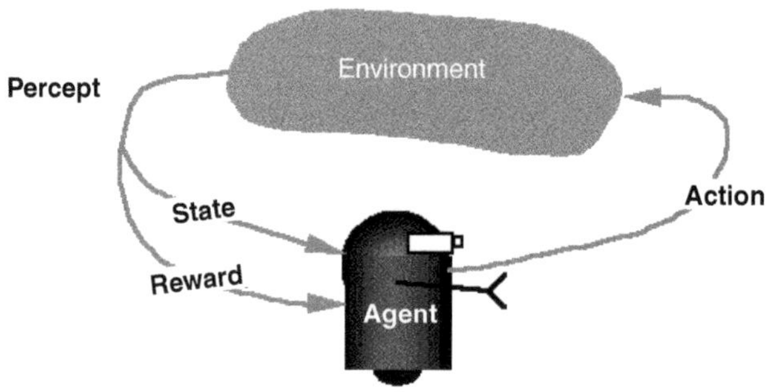

Utility learning

★ *Agent learns utility function.*

★ *Selects actions that maximise expected utility*:

- Disadvantage: must have (or learn) model of environment.
- Need to know where actions lead in order to evaluate actions and make decision.
- Advantage: uses "deeper" knowledge about domain.

Q-learning

★ Agent learns action-value function.

★ Expected utility of taking action in given state.

★ Advantage: no model required.

★ Disadvantage: shallow knowledge.

★ Cannot look ahead & can restrict ability to learn.

Passive learning v.s. Active learning

★ Passive learning:

- The agent simply watches the world going by and tries to learn the utilities of being in various states.

★ Active learning:

- The agent not simply watches, but also acts.

5

Expert Systems

5.1. Expert Systems

An expert system is a computer program that represents and reasons with knowledge of some specialist subject with a view to solving problems or giving advice. To solve expert-level problems, expert systems will need efficient access to a substantial domain knowledge base, and a reasoning mechanism to apply the knowledge to the problems they are given.

Usually they will also need to be able to explain, to the users who rely on them, how they have reached their decisions. They will generally build upon the ideas of knowledge representation, production rules, search, and so on, that we have already covered.

Often we use an expert system shell which is an existing knowledge independent framework into which domain knowledge can be inserted to produce a working expert system. We can thus avoid having to program each new system from scratch.

5.2. Typical Tasks for Expert Systems

There are no fundamental limits on what problem domains an expert system can be built to deal with. Some typical existing expert system tasks include:

1. The interpretation of data, such as sonar data or geophysical measurements.
2. Diagnosis of malfunctions, such as equipment faults or human diseases.
3. Structural analysis or configuration of complex objects, such as chemical compounds or computer systems.
4. Planning sequences of actions, such as might be performed by robots.
5. Predicting the future, such as weather, share prices, exchange rates.

However, these days, "conventional" computer systems can also do some of these things

5.3. Characteristics of Expert Systems

Expert systems can be distinguished from conventional computer systems in that:

1. They simulate human reasoning about the problem domain, rather than simulating the domain itself.
2. They perform reasoning over representations of human knowledge, in addition to doing numerical calculations or data retrieval. They have corresponding distinct modules referred to as the inference engine and the knowledge base.
3. Problems tend to be solved using heuristics (rules of thumb) or approximate methods or probabilistic methods which, unlike algorithmic solutions, are not guaranteed to result in a correct or optimal solution.
4. They usually have to provide explanations and justifications of their solutions or recommendations in order to convince the user that their reasoning is correct.

Note that the term Intelligent Knowledge Based System (IKBS) is sometimes used as a synonym for Expert System.

The Architecture of Expert Systems

The process of building expert systems is often called knowledge engineering. The knowledge engineer is involved with all components of an expert system:

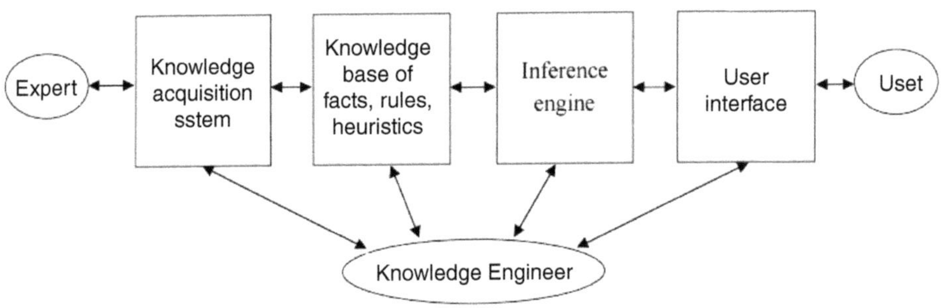

Building expert systems is generally an iterative process. The components and their interaction will be refined over the course of numerous meetings of the knowledge engineer with the experts and users. We shall look in turn at the various components.

5.3.1. Knowledge Acquisition

The knowledge acquisition component allows the expert to enter their knowledge or expertise into the expert system, and to refine it later as and when required.

Historically, the knowledge engineer played a major role in this process, but automated systems that allow the expert to interact directly with the system are becoming increasingly common.

The knowledge acquisition process is usually comprised of three principal stages:

1. Knowledge elicitation is the interaction between the expert and the knowledge Engineer/program to elicit the expert knowledge in some systematic way.
2. The knowledge thus obtained is usually stored in some form of human friendly intermediate representation.
3. The intermediate representation of the knowledge is then compiled into an executable form (*e.g.*, production rules) that the inference engine can process.

In practice, much iteration through these three stages is usually required!

5.3.2. Knowledge Elicitation

The knowledge elicitation process itself usually consists of several stages:

1. Find as much as possible about the problem and domain from books, manuals, etc. In particular, become familiar with any specialist terminology and jargon.
2. Try to characterize the types of reasoning and problem solving tasks that the system will be required to perform.
3. Find an expert (or set of experts) that is willing to collaborate on the project. Sometimes experts are frightened of being replaced by a computer system!

5.4. Meta Knowledge

Knowledge about knowledge:

* Meta knowledge can be simply defined as knowledge about knowledge.
* Meta knowledge is knowledge about the use and control of domain knowledge in an expert system.

5.4.1. Roles in Expert System Development

Three fundamental roles in building expert systems are:

1. Expert - Successful ES systems depend on the experience and application of knowledge that the people can bring to it during its development. Large systems generally require multiple experts.
2. Knowledge engineer - The knowledge engineer has a dual task. This person should be able to elicit knowledge from the expert, gradually gaining an understanding of an area of expertise. Intelligence, tact, empathy, and proficiency in specific techniques of knowledge acquisition are all required of a knowledge engineer.

Knowledge-acquisition techniques include conducting interviews with varying degrees of structure, protocol analysis, observation of experts at work, and analysis of cases.

On the other hand, the knowledge engineer must also select a tool appropriate for the project and use it to represent the knowledge with the application of the knowledge acquisition facility.

5.5. Typical Expert System

A problem-domain-specific knowledge base that stores the encoded knowledge to support one problem domain such as diagnosing why a car won't start. In a rule-based expert system, the knowledge base includes the if-then rules and additional specifications that control the course of the interview.

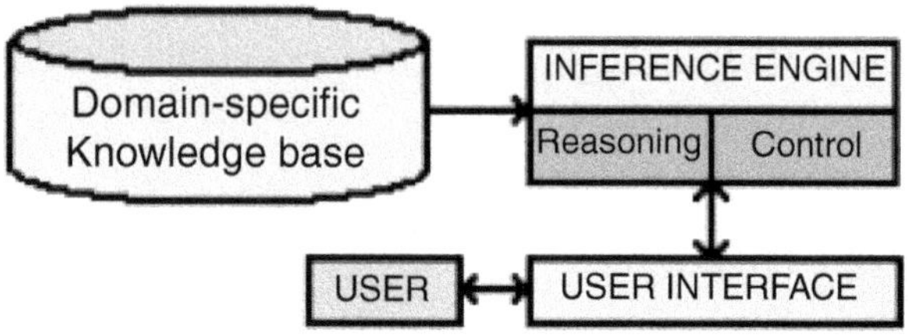

An inference engine a set of rules for making deductions from the data and that implements the reasoning mechanism and controls the interview process. The inference engine might be generalized so that the same software is able to process many different knowledge bases.

The user interface requests information from the user and outputs intermediate and final results. In some expert systems, input is acquired from additional sources such as data bases and sensors.

5.5.1. MYCIN

Tasks and Domain

* Disease DIAGNOSIS and Therapy SELECTION
* Advice for non-expert physicians with time considerations and incomplete evidence on:
    - Bacterial infections of the blood.
    - Expanded to meningitis and other ailments.

System Goals

* Utility:

  Be useful, to attract assistance of experts
    - Demonstrate competence.
    - Fulfill domain need (*i.e.* penicillin).
* Flexibility:
    - Domain is complex, variety of knowledge types.
    - Medical knowledge rapidly evolves, must be easy to maintain K.B.
* Interactive Dialogue:
    - Provide coherent explanations (symbolic reasoning paradigm).
    - Allow for real-time K.B. updates by experts.

★ Fast and Easy:

- Meet time constraints of the medical field.

5.6. DART

The Dynamic Analysis and Replanning Tool, commonly abbreviated to DART, is an artificial intelligence program used by the U.S. military to optimize and schedule the transportation of supplies or personnel and solve other logistical problems.

DART uses intelligent agents to aid decision support systems located at the U.S. Transportation and European Commands. It integrates a set of intelligent data processing agents and database management systems to give planners the ability to rapidly evaluate plans for logistical feasibility. By automating evaluation of these processes DART decreases the cost and time required to implement decisions. DART achieved logistical solutions that surprised many military planners. Introduced in 1991, DART had by 1995 offset the monetary equivalent of all funds DARPA had channel into AI research for the previous 30 years combined.

Development and Introduction

DARPA funded the MITRE Corporation and Carnegie Mellon University to analyze the feasibility of several intelligent planning systems. In November 1989, a demonstration named

The Proud Eagle Exercise indicated many inadequacies and bottlenecks within military support systems. In July, DART was previewed to the military by BBN Systems and Technologies and the ISX Corporation (now part of Lockheed Martin Advanced Technology Laboratories) in conjunction with the CHAPTERed States Air Force Rome Laboratory. It was proposed in November 1990, with the military immediately demanding that a prototype be developed for testing. Eight weeks later, a hasty but working prototype was introduced in 1991 to the USTRANSCOM at the beginning of Operation Desert Storm during the Gulf War.

Impact

Directly following its launch, DART solved several logistical nightmares, saving the military millions of dollars. Military planners were aware of the tremendous obstacles facing moving military assets from bases in Europe to prepared bases in Saudi Arabia, in preparation for Desert Storm. DART quickly proved its value by improving upon existing plans of the U.S. military. What surprised many observers were DART's ability to adapt plans rapidly in a crisis environment.

Expert Systems Shells

Initially each expert system is build from scratch (LISP). Systems are constructed as a set of declarative representations (mostly rules) combined with an interpreter for those representations. It helps to separate the interpreter from domain-specific knowledge and to create a system that could be used construct new expert system by adding new knowledge corresponding to the new problem domain. The resulting interpreters are called shells. Example of shells is EMYCIN (for Empty MYCIN derived from MYCIN).

Shells – A shell is nothing but an expert system without knowledge base. A shell provides the developers with knowledge acquisition, inference engine, user interface, and explanation facility.

For example, few shells are given below:

- ✯ Java Expert System Shell (JESS) that provides fully developed Java API for creating an expert system.
- ✯ Vidwan, a shell developed at the National Centre for Software Technology, Mumbai in 1993. It enables knowledge encoding in the form of IF-THEN rules.

Shells provide greater flexibility in representing knowledge and in reasoning than MYCIN. They support rules, frames, truth maintenance systems and a variety of other reasoning mechanisms.

5.7. AI Application

Finance

Banks use artificial intelligence systems to organize operations, invest into stocks, and manage properties. In August 2001, robots beat humans in a simulated financial trading competition.

Medicine

A medical clinic can use AI systems to organize bed schedules, make a staff rotation, and provide medical information. AI has also application in fields of cardiology (CRG), neurology (MRI), embryology (sonography), complex operations of internal organs, etc.

Social Sites

Facebook wants to analyze the way people communicate with one another so that it can add new features to its services or even automatically remove offensive posts that might occur when a high-profile celebrity like Angelina Joli posts a selfie.

Robotics

Robots are physical agents that perform tasks by manipulating the physical world. Robots are manufactured as hardware. The control of robot is AI (in the form of software agent) that reads data from the sensors decides what to do next and then directs the effectors to act in the physical world.

Heavy Industries

Huge machines involve risk in their manual maintenance and working. So it becomes necessary part to have an automated AI agent in their operation.

Education

AI researchers have created many tools to solve the most difficult problems in computer science.

5.8. Language Models

Formal grammars (*e.g.* regular, context free) give a hard "binary" model of the legal sentences in a language. For NLP, a probabilistic model of a language that gives a probability that a string is a member of a language is more useful.

To specify a correct probability distribution, the probability of all sentences in a language must sum to 1.4.

★ Uses of Language Models:

- Speech recognition – "I ate a cherry" is a more likely sentence than "Eye eight uh Jerry"
- OCR & Handwriting recognition – More probable sentences are more likely correct readings.
- Machine translation – More likely sentences are probably better translations.
- Generation – More likely sentences are probably better NL generations.
- Context sensitive spelling correction

Completion Prediction

A language model also supports predicting the completion of a sentence.

N-Gram Word Models

We can have n-gram models over sequences of words, characters, syllables or other CHAPTERs. Estimate probability of each word given prior context.

An N-gram model uses only N 1 words of prior context.

– Unigram: P(phone)

– Bigram: P(phone I cell)

– Trigram: P(phone I your cell)

The Markov assumption is the presumption that the future behavior of a dynamical system only depends on its recent history. In particular, in a kth-order Markov model, the next state only depends on the *k* most recent states, therefore an N-gram model is a (N 1)-order Markov model.

N-Gram Model Formulas

★ Word sequences w1n w1... wn.

★ Chain rule of probability n n 1 P(w) 2 1 n 1 1 P(w1) P(w2 I w1) P(w3 I w)... P(wn I w.

★ Bigram approximation P(wk I w1k 1)) k 1 n n 1 P(w) P (wk I wk 1) k 1 .

★ N-gram approximation n n 1 k P(wk I wk P(w) 1 N 1) k 1 8.

Estimating Probabilities

★ N-gram conditional probabilities can be estimated from raw text based on the relative frequency of word sequences. Bigram: N-gram: C (wn 1wn) C (wn 1) P (wn | wn 1) n P(wn | wn 1 N 1) n C (wn 1 1wn) N n C (wn 1 1) N

★ To have a consistent probabilistic model, append a unique start (<s>) and end (</s>) symbol to every sentence and treat these as additional words.

N-gram Character Models

★ One of the simplest language models: P(c1N).

★ Language identification: given the text determine which language it is written in.

★ Build a trigram character model of each candidate language: P(ci | ci 2 : i 1 , l).

Smoothing

In practice, parameters are smoothed to reassign some probability mass to unseen events. – Adding probability mass to unseen events requires removing it from seen ones (discounting) in order to maintain a joint distribution that sums to 1.16.

★ Laplace (Add-One) Smoothing
  - The simplest type of smoothing. In the lack of further information, if a random Boolean variable X has been false in all n observations so far then the estimate for P(X=true) should be 1/(n+2).

★ Advanced Smoothing
  - Many advanced techniques have been developed to improve smoothing for language models.

★ Model Combination
  - As N increases, the power (expressiveness) of an N-gram model increases, but the ability to estimate accurate parameters from sparse data decreases (*i.e.*, the smoothing problem gets worse).

★ Interpolation
  - Linearly combine estimates of N-gram models of increasing order. Interpolated Trigram Model: ^ P(wn | wn 2, wn 1) 3 Where: (wn | wn 2, wn 1) i 2 P(wn | wn 1) 1 P(wn)

5.9. Information Retrieval

Information retrieval deals with the storage and representation of knowledge and the retrieval of information relevant for a specific user problem. The information seeker formulates a query trying to describe his information need. The query is compared to document representations. The representations of documents and queries are typically matched by a similarity function such as the Cosine or the Dice coefficient. The most similar documents are presented to the users who can evaluate the relevance with respect to their problem.

Indexing usually consists of the several phases. After word segmentation, stopwords are removed. These common words like articles or prepositions contain little meaning by themselves and are ignored in the document representation. Second, word forms are transformed into their basic form, the stem. During the stemming phase, *e.g.* houses would be transformed into house. For the document representation, different word forms are usually not necessary. The importance of a word for a document can be different. Some words better describe the content of a document than others.

In multimedia retrieval, the context is essential for the selection of a form of query and document representation. Different media representations may be matched against each other or transformations may become necessary (*e.g.* to match terms against pictures or spoken language utterances against documents in written text).

As information retrieval needs to deal with vague knowledge, exact processing methods are not appropriate. Vague retrieval models like the probabilistic model are more suitable. Within these models, terms are provided with weights corresponding to their importance for a document. These weights mirror different levels of relevance.

The result of current information retrieval systems are usually sorted lists of documents where the top results are more likely to be relevant according to the system. In some approaches, the user can judge the documents returned to him and tell the systems which ones are relevant for him. The system then resorts the result set. Documents which contain many of the words present in the relevant documents are ranked higher. This relevance feedback process is known to greatly improve the performance. Relevance feedback is also an interesting application for machine learning.

5.10. Information Extraction (IE)

- ✯ Identify specific pieces of information (data) in a unstructured or semi-structured textual document.
- ✯ Transform unstructured information in a corpus of documents or web pages into a structured database.
- ✯ Applied to different types of text:
  - Newspaper articles
  - Web pages
  - Scientific articles
  - Newsgroup messages
  - Classified ads
  - Medical notes
  - Wikipedia (info boxes).

Information Extraction vs. NLP?

- ✯ Information extraction is attempting to find *some* of the structure and meaning in the hopefully template driven web pages.

- ★ As IE becomes more ambitious and text becomes more free form, then ultimately we have IE becoming equal to NLP.
- ★ Web does give one particular boost to NLP
    - *Massive* corporation

5.11. Natural Language Processing

The field of Natural Language Processing (NLP) goes back to the 1950s. Yet it is still of crucial importance today for the new information society. Its goal is to process natural language texts, either for analysing existing texts or generating new ones, for the purpose of achieving human-like language processing for a range of tasks or applications. These applications, regrouped under the term **language engineering",** include machine translation, question answering, information retrieval, information extraction, data mining, reading and writing aid, and many others.

From a more research-oriented point of view, empirical linguistics and digital humanities can also be viewed as application domains of NLP. NLP is a transdisciplinary domain; it requires an expertise in formal and descriptive linguistics (to develop linguistic models of human languages), in computer science and algorithmics (to design and develop efficient programs that can deal with such models) and in applied mathematics (to automatically acquire linguistic or general knowledge).

Processing natural language texts is a difficult task, in particular because of the large amount of ambiguity in natural language, the specificities of individual languages and dialects and because many users do not necessarily conform to grammatical and spelling conventions, when such conventions exist.

The first decades of NLP mostly focused on symbolic approaches, also contributing major notions to Computer Science, especially in formal grammar theory and parsing techniques. Linguistic knowledge was mostly encoded in the form of manually developed grammars and lexical databases.

Over the last two decades statistical and machine learning based approaches have greatly renewed the field, bringing annotated corpora to centre stage, and significantly improving the state of the art. However symbolic approaches retain specific advantages, and best results are often obtained when leveraging all types of resources including linguistic knowledge within hybrid systems coupling symbolic and statistical techniques.

Natural Language Processing Applications

- ★ Speech technologies.
- ★ Automatic speech recognition (ASR)
- ★ Text-to-speech synthesis (TTS)
- ★ Dialog systems

Language Processing Technologies

- ★ Machine translation

- ★ Information extraction
- ★ Web search, question answering
- ★ Text classification, spam filtering, etc.

5.12. Machine Translation

- ★ Translate text from one language to another
- ★ Recombines fragments of example translations

Challenges

- ★ What fragments? [learning to translate]
- ★ How to make efficient? [fast translation search]

Fundamental goal: analyze and process human language, broadly, robustly, accurately.

End systems that we want to build

Ambitious: speech recognition, machine translation, information extraction, dialog interfaces, question answering...

Modest: spelling correction, text categorization, language recognition, genre classification.

Types

- ★ **Rough translation:** gives the main point but contains errors.
- ★ **Pre-edited translation:** original text written in constrained language easier to translate automatically.
- ★ **Restricted-source translation**: fully automatic but only on technical content as, *e.g.* weather forecast.

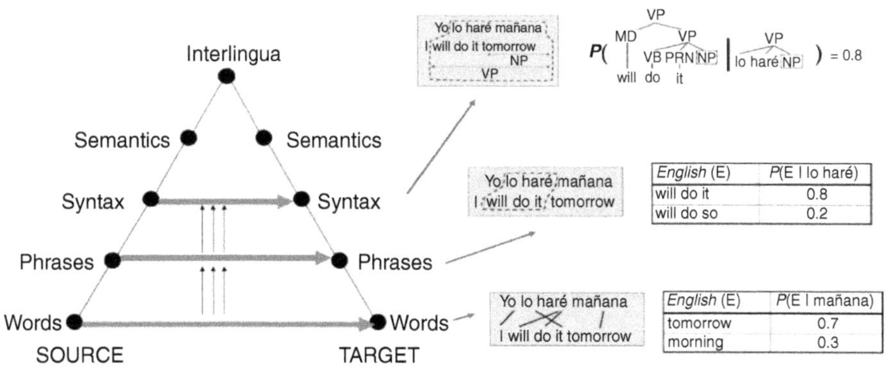

Machine Translation Systems

Very simplified, there are three types of machine translation:

Statistical machine translation (SMT) learn relational dependencies of features such as grams, lemmas, etc. Requires large data sets.

Example: google translate. Relatively easy to implement.

Rule-based machine translation (RBMT) use grammatical rules and language constructions to analyze syntax and semantics. Use moderate size data sets. Long development time and expertise.

Hybrid machine translation either construct from RBMT and use SMT to post-process and optimize the result. Or use grammatical rules to derive further features to then be fed in the statistical learning machine. New direction of research.

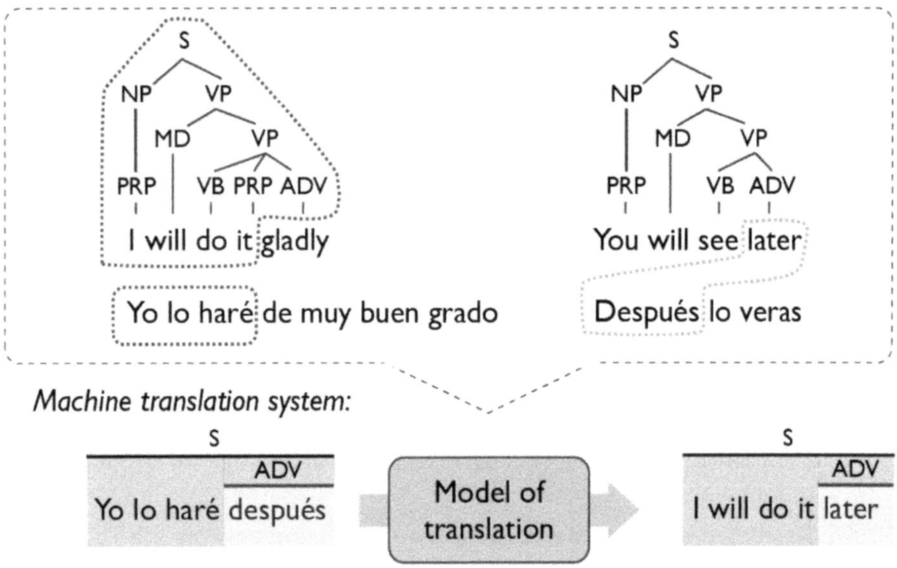

Grammars Machine Translation

Grammars: set of rules (from left to right) that describe how to form strings from the language's alphabet that are valid according to the language's syntax (Language generator).

Parsing is the process of recognizing a string in natural languages by breaking it down to a set of symbols and analyzing each one against the grammar of the language, *i.e.*, determining whether the string belongs to the language or is grammatically incorrect. The result is a parse tree.

- Context free grammars (see http://en.wikipedia.org/wiki/Chomsky_hierarchy).
- Probabilistic context free grammars.
- Lexicalized probabilistic context free grammars.

5.13. Speech Recognition

Digitalizing Speech

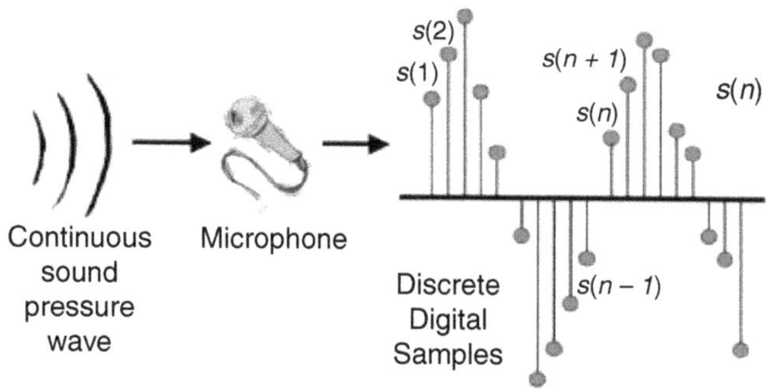

Speech Input is an Acoustic Wave Form

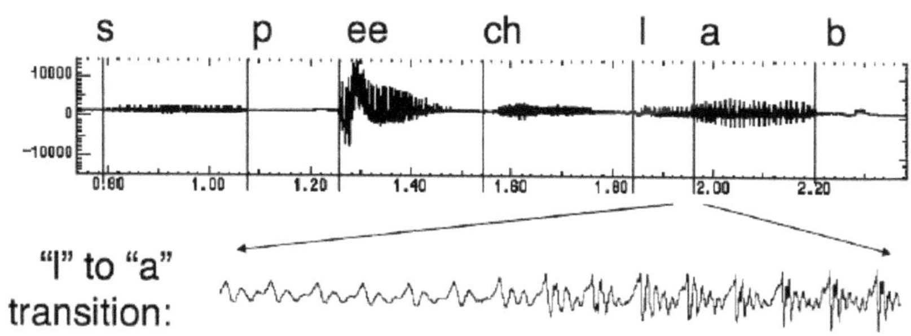

Spectral Analysis

- Frequency gives pitch; amplitude gives volume
  - sampling at ~8 kHz phone, ~16 kHz mic (kHz=1000 cycles/sec)

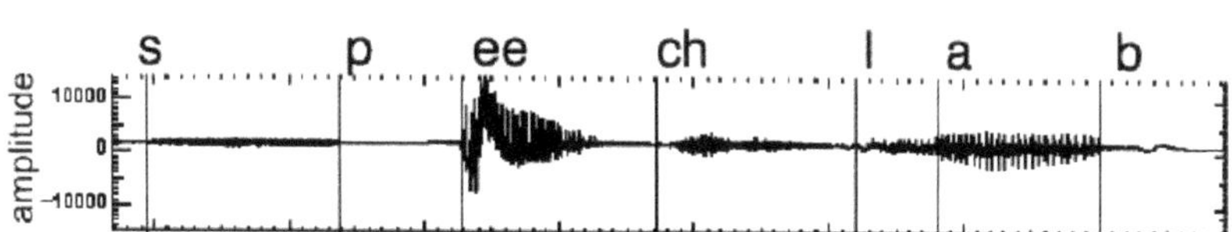

- Fourier transform of wave displayed as a spectrogram
  - darkness indicates energy at each frequency

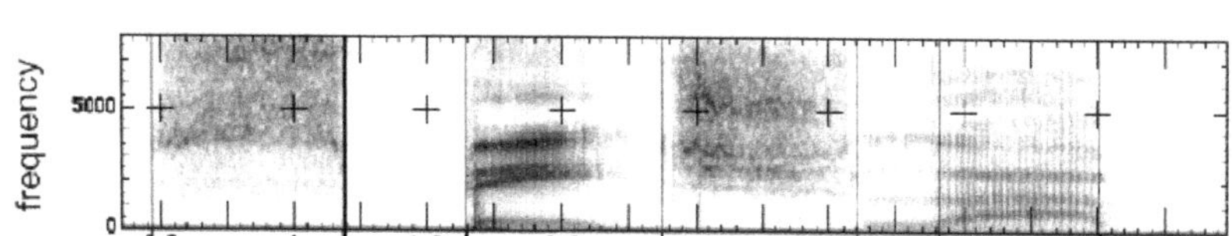

HMM for Speech

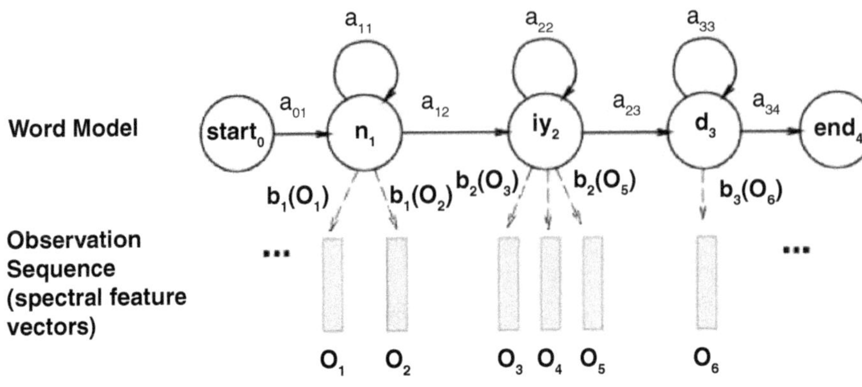

5.14. Robotics in AI

Building on the advances made in mechatronics, electrical engineering and computing, robotics is developing increasingly sophisticated sensorimotor functions that give machines the ability to adapt to their ever-changing environment.

Until now, the system of industrial production was organized around the machine; it is calibrated according to its environment and tolerated minimal variations.

Today, it can be integrated more easily into an existing environment. The autonomy of a robot in an environment can be subdivided into perceiving, planning and execution (manipulating, navigating, collaborating).

The main idea of converging AI and Robotics is to try to optimise its level of autonomy through learning. This level of intelligence can be measured as the capacity of predicting the future, either in planning a task, or in interacting (either by manipulating or navigating) with the world. Robots with intelligence have been attempted many times.

Although creating a system exhibiting human-like intelligence remains elusive, robots that can perform specialized autonomous tasks, such as

- ★ Driving a vehicle,
- ★ Flying in natural and Man-made environments,
- ★ Swimming,
- ★ Carrying boxes and material in different terrains,
- ★ Pick up objects and put them down do exist today.

5.14.1. Perception

Another important application of AI in robotics is for the task of perception.

Robots can sense the environment by means of integrated sensors or computer vision.

In the last decade, computer systems have improved the quality of both sensing and vision.

Perception is not only important for planning but also for creating an artificial sense of self-awareness in the robot.

This permits supporting interactions with the robot with other entities in the same environment.

This discipline is known as social robotics. It covers two broad domains: human-robot interactions (HCI) and cognitive robotics.

The vision of HCI it to improve the robotic perception of humans such as in understanding activities , emotions , non-verbal communications and in being able to navigate an environment along with humans.

The field of cognitive robotics focuses on providing robots with the autonomous capacity of learning and acquiring knowledge from sophisticated levels of perception based on imitation and experience.

It aims at mimicking the human cognitive system, which regulates the process of acquiring knowledge and understanding, through experience and sensorisation.

In cognitive robotics, there are also models that incorporate motivation and curiosity to improve the quality and speed of knowledge acquisition through learning.

AI has continued beating all records and overcoming many challenges that were unthinkable less than a decade ago. The combination of these advances will continue to reshape our understanding about robotic intelligence in many new domains.

Milestone in Robotics and AI.

5.15. Planning

The task of coming up with a sequence of actions that will achieve a goal is called **planning**.

We have seen two examples of planning agents so far: the search-based problem-solving agent and the logical planning agent.

- ✯ We studied how to take actions in the world (search).
- ✯ We studied how to represent objects, relations, etc. (logic).

5.15.1. The Planning Problem

Let us consider what can happen when an ordinary problem-solving agent using standard search algorithms—depth-first, A_, and so on—comes up against large, real-world problems. That will help us design better planning agents. The most obvious difficulty is that the problem-solving agent can be overwhelmed by irrelevant actions.

In recent years, it has become clear that STRIPS is insufficiently expressive for some real domains. As a result, many language variants have been developed. The Action Description Language or **ADL**, by comparing it with the basic STRIPS language. In ADL, the Fly action could be written as:

Action(Fly(p : Plane; from : Airport; to : Airport)

PRECOND : At (p; from) ^ (from 6= to)

EFFECT : At (p; from) ^ At (p; to))

The notation p : Plane in the parameter list is an abbreviation for Plane(p) in the precondition; this adds no expressive power, but can be easier to read. (It also cuts down on the number of possible propositional actions that can be constructed.) The precondition (from 6= to) expresses the fact that a flight cannot be made from an airport to itself. This could not be expressed succinctly in STRIPS.

The various planning formalisms used in AI have been systematized within a standard syntax called the the Planning Domain Definition Language, or PDDL. This language allows researchers to exchange becnchmark problems and compare results.

```
Init(At(C1; SFO) ^ At(C2; JFK) ^ At(P1; SFO) ^ At(P2; JFK)
^ Cargo(C1) ^ Cargo(C2) ^ Plane(P1) ^ Plane(P2)
^ Airport(JFK) ^ Airport(SFO))
Goal (At(C1; JFK) ^ At(C2; SFO))
Action(Load(c; p; a);
PRECOND: At(c; a) ^ At(p; a) ^ Cargo(c) ^ Plane(p) ^ Airport(a)
EFFECT: : At(c; a) ^ In(c; p))
Action(Unload(c; p; a);
PRECOND: In(c; p) ^ At (p; a) ^ Cargo(c) ^ Plane(p) ^ Airport(a)
EFFECT: At(c; a) ^ : In(c; p))
Action(Fly(p; from; to);
PRECOND: At(p; from) ^ Plane(p) ^ Airport(from) ^ Airport(to)
EFFECT: : At(p; from) ^ At(p; to))
```

A STRIPS problem involving transportation of air cargo between airports.

The STRIPS and ADL notations are adequate for many real domains. The subsections that follow show some simple examples. There are still some significant restrictions, however.

The most obvious is that they cannot represent in a natural way the **ramifications** of actions. For example, if there are people, packages, or dust motes in the airplane, then they too change location when the plane flies. We can represent these changes as the direct effects of flying, whereas it seems more natural to represent the location of the plane's contents as a logical consequence of the location of the plane.

Example: The Spare Tire Problem

Consider the problem of changing a flat tire. More precisely, the goal is to have a good spare tire properly mounted onto the car's axle, where the initial state has a flat tire on the axle and a good spare tire in the trunk. To keep it simple, our version of the problem is a very abstract one, with no sticky lug nuts or other complications.

There are just four actions:

- removing the spare from the trunk,
- removing the flat tire from the axle,
- putting the spare on the axle, and
- Leaving the car unattended overnight.

We assume that the car is in a particularly bad neighbourhood, so that the effect of leaving it overnight is that the tires disappear.

```
Init(At(Flat; Axle) ^ At(Spare;Trunk))
Goal (At(Spare; Axle))
Action(Remove(Spare;Trunk),
PRECOND: At (Spare;Trunk)
EFFECT: : At(Spare;Trunk) ^ At(Spare; Ground))
Action(Remove(Flat; Axle),
PRECOND: At (Flat; Axle)
EFFECT: : At(Flat; Axle) ^ At(Flat; Ground))
Action(PutOn(Spare; Axle),
PRECOND: At(Spare; Ground) ^ : At(Flat; Axle)
EFFECT: : At(Spare; Ground) ^ At(Spare; Axle))
Action(LeaveOvernight,
PRECOND:
EFFECT: : At(Spare; Ground) ^ : At(Spare; Axle) ^ : At(Spare;Trunk)
^ : At(Flat; Ground) ^ : At(Flat; Axle))
```

The Simple Spare Tire problem

It goes beyond STRIPS in that it uses a negated precondition, :At (Flat; Axle), for the PutOn(Spare; Axle) action.

Weak and Strong AI

When defining the capacity of AI, this is frequently categorised in terms of weak or strong AI. Weak AI (narrow AI) is one intended to reproduce an observed behaviour as accurately as possible. It can carry out a task for which they have been precision-trained. Such AI systems can become extremely efficient in their own field but lack generalisational ability. Most existing intelligent systems that use machine learning, pattern recognition, data mining or natural language processing are examples of weak AI. Intelligent systems, powered with weak AI include recommender systems, spam filters, self-driving cars, and industrial robots.

Strong AI is usually described as an intelligent system endowed with real consciousness and is able to think and reason in the same way as a human being. A strong AI can, not only assimilate information like a weak AI, but also modify its own functioning, *i.e.* is able to autonomously reprogram the AI to perform general intelligent tasks. These processes are regulated by human-like cognitive abilities including consciousness, sentience, sapience and self-awareness. Efforts intending to generate a strong AI have focused on whole brain simulations, however this approach has received criticism, as intelligence cannot be simply explained as a biological process emanating from a single organ but is a complex coalescence of effects and interactions between the intelligent being and its environment, encompassing a series of diverse ways via interlinked biological process.

Summary

Early expert system shells provide mechanisms for knowledge representation, reasoning and explanation. Later these tools provide knowledge acquisition. Still expert system shells need to integrate with other programs easily. Expert systems cannot operate in a vacuum. The shells must provide an easy-to-use interface between an expert system written with the shell and programming environment. Applications of AI and current trends in AI enhance the technology in the environment.

Index

G

H

I

K

L

M

N

O

P

www.ingramcontent.com/pod-product-compliance
Ingram Content Group UK Ltd.
Pitfield, Milton Keynes, MK11 3LW, UK
UKHW020047200726
13845UKWH00044B/31

9 789390 371792